AN UNDERWATER GUIDE TO INDONESIA

Photo credit: All photos taken in Indonesia by the author.

Published in North America by
University of Hawai'i Press
2840 Kolowalu Street
Honolulu, Hawai'i 96822

First Published in Singapore by
Times Editions
An imprint of Times Media Private Limited
A member of the Times Publishing Group
Times Centre, 1 New Industrial Road
Singapore 536196
Tel: (65) 2848844 Fax: (65) 2854871
Email: te@corp.tpl.com.sg
online bookstore: http://www.timesone.com.sg/te

Printed in Singapore

Library of Congress Cataloging-in-Publication Data

A catalogue record for this book has been requested

ISBN 0-8248-2368-0

AN UNDERWATER GUIDE TO
INDONESIA

R. Charles Anderson

University of Hawai'i Press
Honolulu

Contents

*The strikingly patterned zebra moray, **Gymnomuraena zebra**, is normally only seen at night.*

Foreword

Indonesia is a country with far more sea than land. In recognition of the importance of the sea in their lives, Indonesians refer to their country as *tanah air kita*, 'our land and water'. The marine life of Indonesia is second to none in terms of both numbers of species and diversity of forms. There are more species of fish, more species of coral, more of almost all marine life groups in Indonesia than in any other country.

The aim of this book is to provide an introduction to this astounding array of marine life. It does not pretend to be anywhere near comprehensive. Rather, it is intended to whet the appetite with a selection of the delights found underwater in Indonesia. I have concentrated on the coral reefs, which are the areas most frequented by divers and snorkellers. However, other types of marine habitat are well worth exploring, for they too host a multitude of interesting creatures. In fact, one key to finding a particular species is to look in its particular habitat. Another is to look hard. This may be stating the obvious but too many divers and snor-

kellers pass their time in a daze, swimming blindly from one big object to the next. They miss out on the myriad beautiful and fascinating small creatures that are there on show for anyone who takes the time to look. Slow down, look systematically, and all will be revealed!

I am most grateful to the many people who have assisted me in Indonesia in various ways over the years, particularly with diving, from West Sumatra to Maluku. Particular thanks to Drs. Frans Seda and the staff of Sao Wisata Resort, Maumere, Flores; Loky Herlambang and the staff of Nusantara Diving Centre, Manado, Sulawesi; Pak Nawawi and the staff of Derawan Island Resort, East Kalimantan; Sony and Carol and the staff of Ambon Dive Centre, Ambon; the staff of Dive Paradise Tulamben, Bali. Dr. Ted Pietsch kindly confirmed frogfish identifications, and Dr. Vic Springer assisted with blennies. RGB Color have provided fast and efficient film processing during all too brief stopovers in Singapore. My parents have given much support over the years, for which I am truly grateful. To my lovely wife and diving buddy, Susan, a special 'thank you'.

R. Charles Anderson

This crinoid clingfish, **Discotrema echinophila**, *makes its home within the protective arms of a featherstar.*

Indonesia

A country of superlatives, Indonesia is the world's largest archipelago, with an astonishing 17,500 islands (this is only a best estimate, with the true number having been put at anything from 13,500 to 18,500). It is home to three of the world's five largest islands — New Guinea (which includes Irian Jaya), Borneo (which includes Kalimantan) and Sumatra are the second, third and fifth largest of all islands (while Greenland and Madagascar are in first and fourth places). Indonesia has over 200,000 km of coastline, which is equivalent to over five times the distance around the Earth at the equator (although another estimate puts the coastline at 'only' 81,000 km).

Indonesia is also superabundantly endowed when it comes to marine life. Marine scientists divide the oceans up into various regions according to the distributions of marine life. The largest of all is the deep-sea. But in the surface waters, by far the largest marine life region is the Indo-Pacific. This is a vast swathe of tropical ocean that stretches halfway around the globe, from the

Snorkelling in the Indian Ocean off Pulau Sikuai, West Sumatra.

East African coast to the furthest islands of Polynesia. Not only is it the largest marine life region, but it is also by far the richest. Within this enormous region, the greatest diversity is found in the central area encompassed by Malaysia, Indonesia, southern Philippines and Papua New Guinea. This area is sometimes known as the Indo-Malayan triangle. Indonesia forms the largest and most central part of it — here is the greatest marine biodiversity of all.

Indonesian marine life is not only rich on the global scale, but also at the local level. In the middle of the 19th century, nearly 800 species of fishes were recorded from Ambon Bay alone by Pieter Bleeker (1819-78), and that was without the aid of modern diving equipment!

Bleeker was a medical officer in the Dutch colonial army, and the greatest student of Indonesian marine life. He first arrived in what is now Indonesia in 1842, and soon started to collect fishes as a hobby. Over the next 30 years he amassed an enormous collection of Indonesian fishes. He described hundreds of them in a mountain of scientific publications,

Ambon Bay (Teluk Ambon) is a great sea arm roughly 30 km long, which almost bisects the island. Pieter Bleeker studied the rich fish fauna here in the middle of the 19th century. Today the inner bay around Ambon town is badly degraded, but the outer parts of the bay offer some superb 'muck diving'.

culminating in his masterwork, the nine-volume *Atlas Ichthyologique*. This is still recognized today as the keystone of Indonesian ichthyology.

More recently, Australian ichthyologists have recorded over 1,100 fish species from Maumere Bay in Flores, and they are still counting. No one knows the total number of fish species to be found in Indonesia, but it must be over 3,000. For comparison, just 1,250 species of fish are known from all European waters.

Turning to corals, over 450 species of reef corals are known from Indonesia so far, and again new species

are still being added to the total. For comparison, the whole of the Caribbean boasts barely 50 reef-building coral species. The list could go on and on. In almost every group of tropical marine organisms studied so far, there are more species in Indonesia than anywhere else.

Patterns of marine biodiversity

Within Indonesia itself, species richness is higher in the east than in the west. For example, eastern Indonesia has 15 species of rabbitfishes (more than any other area in the world), while western Indonesia has 12. If the Indo-Malayan triangle is the centre of Indo-Pacific marine biodiversity, then eastern Indonesia is its very epicentre.

While it is clear that eastern Indonesia has the highest species diversity in the sea, explaining this pattern is something that has exercised the talents of marine scientists for decades.

One explanation is that some parts of Indonesia have been in the same equatorial latitudes for hundreds of millions of years. In contrast, other land masses have moved, by continental drift, from tropical to polar latitudes (or vice versa). While tropical marine species flourished in Indonesia's stable environment, many species elsewhere were driven to extinction. This is not to say that there have been no dramatic changes in Indonesia. The Tethys Sea was an ancient tropical ocean which once extended from Indonesia to the Atlantic. It all but

disappeared about 15 million years ago as a result of continental drift. Recent research in the Togian Islands (sheltered between the sweeping arms of northern and central Sulawesi) suggests that some of the unusual corals there may be relicts of a once more widespread Tethys Sea fauna.

Another factor to be taken into account is that the tropical Indian and Pacific Oceans meet in Indonesia. Thus in the case of fishes, Sumatra, Java and the south coasts of many of the lesser Sunda Islands have a predominantly Indian Ocean fish fauna. In contrast, the rest of Indonesia has a predominantly Pacific Ocean fish fauna. For this reason alone, Indonesia certainly has more marine species than neighbouring countries, which are sometimes touted as having greater marine biodiversity.

Amed on the north coast of Bali offers fine diving and snorkelling. It is overshadowed by the volcanic mountain Gunung Agung.

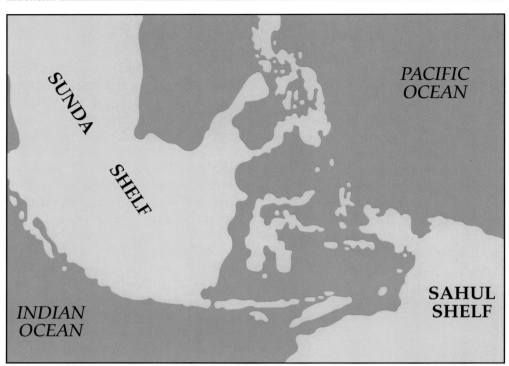

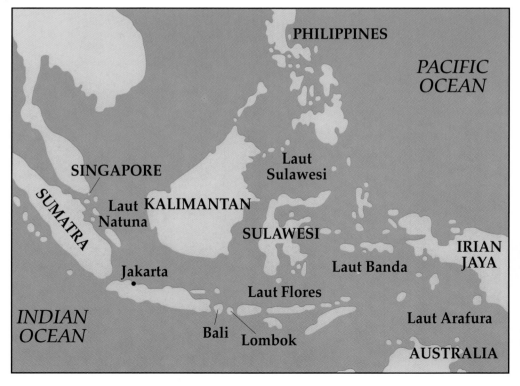

Ice Ages and Sea Levels

The barrier between the Indian and Pacific Oceans presented by the Indonesian Archipelago has at various times in the past been very much more complete than it is now. Over the last 100,000 years there were a series of major glaciations or Ice Ages. So much water was locked up in the polar ice caps during each glaciation that the sea level dropped far below what we have today. At the peak of the last glaciation, some 21,000 years ago, the sea level was about 120 metres lower than it is now.

Much of Indonesia's archipelagic waters are shallower than 120 metres, and so these areas were exposed as dry land during the last glaciation. In western Indonesia between Sumatra and Java on one side and Kalimantan on the other (the Laut Natuna and Laut Java) is the Sunda Shelf. Here the waters are relatively shallow, only occasionally exceeding 50 metres in depth. Similarly, the waters of eastern Indonesia, south of Irian Jaya (the Laut Arafura) lie over the Sahul Shelf and are mostly less than 100 metres deep. In contrast, the central archipelagic waters of the Laut Flores and Laut Banda are very deep. During the Ice Ages, the whole of the Sunda and Sahul Shelves were exposed as dry land. This is how rhinoceroses and tigers managed to get from mainland Asia to Borneo and Bali, but not to Lombok and Sulawesi.

In this way connections between the Indian and Pacific Oceans were greatly reduced during glacial times.

Opposite: Southeast Asia during the height of the last Ice Age (top) and today (bottom).

Widespread populations of marine creatures were split into two. In some cases the two forms diverged sufficiently to warrant being recognized today as separate species. Such species pairs are known as siblings or as geminate species. Indonesia has more marine sibling species pairs than any other country.

So great is the diversity of marine life in Indonesia, that marine scientists are still struggling to catalogue all the species present. Even among relatively well studied groups such as corals and fishes, new species are being discovered on a regular basis. For less well known groups such as sponges

*The saddleback butterflyfish, **Chaetodon falcula** (below, photographed off West Sumatra) is the Indian Ocean species of a sibling pair. In the Pacific it is replaced by the double-saddled butterflyfish, **Chaetodon ulietensis** (bottom, photographed off Flores).*

and worms there must be hundreds if not thousands of new species awaiting discovery. The ecological interactions of all these species, their behaviours and life histories are all but unknown. While professional marine scientists are working on some of these issues, there is still room for the amateur. Any keen diver who picks a group of animals for detailed investigation, or a small area for regular study can expect to make many new discoveries. Such studies are needed now more than ever, because Indonesia's great wealth of marine life is under threat as never before.

Conservation

Indonesia is beset by numerous problems affecting the conservation of marine resources and their sustainable utilization. Pollution and unplanned coastal development are having a major impact on coral reefs, as well as mangrove forests and sea grass meadows. At the same time there is a growing demand for marine resources. Not only does Indonesia have a large and growing population which needs feeding, but also overseas demand for luxury sea food items is burgeoning. Export-oriented fisheries have developed to meet these demands, with catastrophic results for some species. Because high prices are paid for these commodities in Singapore, Taipei, Hong Kong, Tokyo and elsewhere, fishermen will continue to fish even when stocks have been driven to near local extinction as a result of intense fishing pressure.

Seven species of giant clams (*Tridacna* and *Hippopus* species, *kima* in Indonesian) are found in Indonesia. They are highly valued for their meat, i.e. the muscle that pulls the two halves of the shell together. The rest of the living tissue is normally discarded, although for some years their shells were used in Java to produce terrazzo floor tiles. Larger individuals had been harvested sustainably for thousands of years, but exploitation of giant clams was so high in the 1970s and 1980s that they became extinct on many Indonesian reefs. Giant clams have been protected by law since 1987, but some fishing still continues.

Groupers and napoleon wrasses are a delicacy throughout much of East Asia, and a major export industry has developed to meet this demand. Again, some species have been driven to local extinction in some areas. Illegal fishing methods such as the use of cyanide (by snorkellers and divers on crude hookah equipment) also causes considerable damage to corals.

Sea horses are prized in traditional Chinese medicine, and are widely collected for export by fishermen and other coastal villagers. They are becoming increasing rare throughout Indonesia. When I saw a sea horse in shallow sea grass at Bunaken Island off Manado, I foolishly made a big noise about it. Before I realized what was happening, a passing fisherman grabbed the poor beast and disappeared. It was a whole year before I found another one!

Shrimps are a delicacy through-

*Opposite top: Giant clams, **Tridacna**, have been heavily overfished in many areas of Indonesia.*
*Opposite bottom: Vermilion rock cod, **Cephalopholis miniata**, a favourite with both divers and fishermen.*

out much of the world, and shrimp trawlers dragging the rich grounds of the Laut Arafura have wreaked untold damage on the marine habitats there. Meanwhile, the development of prawn farms in other coastal areas has brought devastation to mangroves. Ironically, mangroves are vital nursery grounds for shrimps. The greatest shrimp catches are made offshore from the largest mangrove forests. Destroying coastal mangrove forests to make room for prawn farms is destroying existing small-scale shrimp fisheries.

Divers and snorkellers can play a positive role here. We should all review our own consumption habits. If you eat shark fin soup, do not complain about the lack of sharks at dive sites. If you eat grouper, expect your

holiday reef destinations to be damaged. There is a connection! In addition, if every one of us who visits popular sites breaks just a couple of pieces of coral, there will soon be little left to see. Take real care when in the water not to damage any corals. Learn to recognize the difference between living corals and dead rock in case you have to hold onto something, for example in a current. Do not wear gloves when diving: if you don't know what it is, don't touch it!

Below: Two blue-spotted fantail rays, **Taeniura lymma**, *discarded by fishermen on a reef popular with divers. Dumping unwanted catch at sea is normally of little significance in artisanal fisheries, but it is a major problem with large-scale fisheries, notably shrimp trawling.* **Opposite:** *A fish vendor in Jakarta. Small-scale local fisheries are of vital importance for millions throughout Indonesia, providing both employment and protein.*

Marine Habitats

The coral reef is the marine habitat that first comes to mind for many of us when we think of tropical seas. It is indeed a habitat of major importance, and as such it is dealt with in a separate section below. However, coral reefs are not the only significant marine habitat in Indonesia. They are not even the largest. That honour goes to the open sea.

The open sea

Open sea accounts for just over three quarters of the area under Indonesian jurisdiction, nearly 6 million km². While it is in some respects a desert in comparison to coral reefs, it is still home to many animals, and to important fisheries. When travelling by boat between islands or dives, it is easy to nod off, to pass the time in conversation, or to get lost in a book. But keep a sharp lookout and you are almost sure to be rewarded with sightings of seabirds, dolphins, fish schools or turtles.

Mangrove forests are important coastal ecosystems in many areas of Indonesia. The so-called knee roots are characteristic of **Bruguiera** *mangrove trees, and are a means by which the roots can get oxygen even though the trees are living in anoxic mud.*

Whales too are a real possibility, with 14 species known from Indonesian waters so far. One memorable morning while enjoying a coffee on the beach before a dive at Tulamben on the north coast of Bali, I watched a pair of Bryde's whales (*Balaenoptera edeni*) passing offshore. One of them breached (leapt half out the water) half a dozen times before they disappeared from view. Traditional small-boat whaling still survives in two villages in Nusa Tenggara Timur, where it has been carried out for hundreds of years. Fishermen from Lamakera on Solor catch small numbers of baleen whales each year, while those from Lamalera on Lembata target sperm whales (*Physeter macrocephalus*) and other smaller toothed whales.

Mangroves

Mangroves are trees that are adapted to life in the intertidal zone, and can survive regular flooding by salt water. They form thick forests or mangals along suitable coasts. Mangroves may appear unappealing to the uninitiated, but they are well worth exploring with a local guide.

It is also worth diving in front of mangrove stands, at high tide. Such 'muck dives' usually turn up a wealth of bizarre fishlife.

Indonesia has the largest area of mangroves in the world. Some estimates put the original area of these forests at 42,500 km². However, coastal 'development' has had disastrous consequences for Indonesia's mangroves, with perhaps 50% having been lost nationally. Java has lost roughly 90% of its mangroves. Over half of all that remains is found in remote Irian Jaya. Other important mangrove areas are the coasts of eastern Sumatra, Kalimantan and southwest Sulawesi.

The loss of mangrove forests can have serious consequences for coastal defence. It also eliminates important sources of firewood and timber. Per-haps even more importantly, it can have devastating effects on marine fisheries nearby, because mangrove forests are vital nursery grounds for many commercial fishery species.

Sea grass

Sea grasses are land plants that have returned to the sea. Unlike the seaweeds or algae, sea grasses have true roots and flowers, which are pollinated underwater. Sea grasses form extensive meadows in many bays and lagoons, and they also grow on some Indonesian coral reefs.

Divers and snorkellers often hurry over sea grass beds in their rush to get to the coral reefs further out. This is a shame, for sea grass meadows are home to a surprisingly diverse array of interesting creatures. Look out for lurid green wrasses, herbivorous rabbitfishes, and schools of juvenile

Sea grass meadows are well worth exploring. They are home to a surprising number of animals, such as this pufferfish, **Arothron hispidus.**

fishes. As with mangroves, sea grass meadows are important nursery grounds for many species of commercially important fish. A night snorkel on a sea grass bed will often reveal many animals, such as eels and crabs, that remain buried by day. Sea grass meadows are also important grazing areas for green turtles and dugongs.

*Sponges as well as sea squirts (**Styela complexa**) and mussels (**Brachydontes variabilis**) crowd the roots of mangrove trees just beneath the surface of Kakaban marine lake, East Kalimantan.*

Marine lakes

One special type of habitat that is minute in terms of size, but holds limitless interest is the marine lake. The lake on Kakaban island, off East Kalimantan, was formed when Kakaban atoll was raised about 60 metres by geological forces. The atoll lagoon became an enclosed marine lake with a maximum depth of about 11 metres. It is encircled by the old atoll reef, which now stands like an ancient castle rampart, some 45-60 metres high and thickly forested. Seawater can seep through cracks and crevices to the lake, which therefore does experience tides, albeit reduced in magnitude and delayed by about 3 hours. The lake is surrounded by mangroves and supports a magical array of other inhabitants including marine algae, numerous small fishes and four species of jellyfish.

Even more unusual is Lake Motitoi on Satonda Island, north of Sumbawa. This is a crater lake filled with seawater, but long cut off from the sea. It is nearly 70 metres deep. Its deeper waters are virtually lifeless, but the shoreline is home to most unusual algal reefs known as stromatolites. These are well known from fossils over 600 million years old, but today survive at only three other localities worldwide.

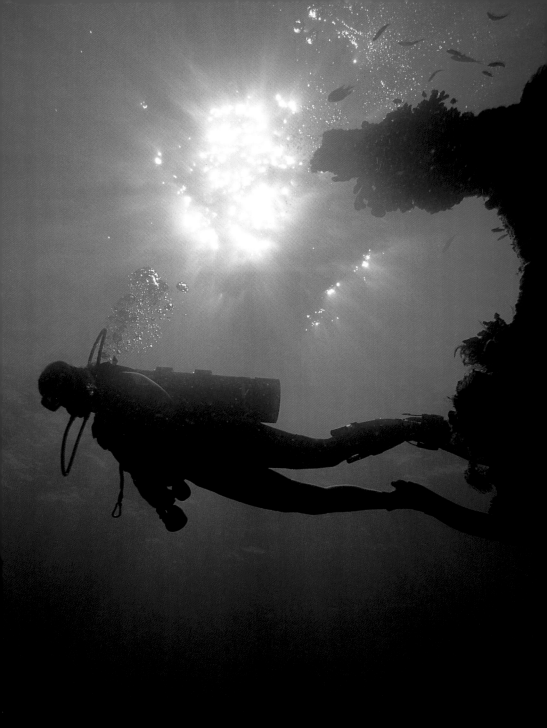

Coral Reefs

Whatever the attractions of mangroves and sea grass beds, coral reefs are undoubtedly the most interesting of all marine habitats for the majority of divers and snorkellers. For one thing, coral reefs are home to a greater variety of species than any other marine habitat. In addition, corals require clear, warm waters to thrive, and these are just the conditions that are so attractive to people.

Indonesia has a huge area of coral reefs, but exactly how much is not certain. The reason for this is that coral reefs are notoriously difficult to measure accurately. Most estimates are based on navigational charts, supplemented by aerial or satellite photography. It is often impossible to calculate the extent of deep reefs from these sources, which are normally not accurate enough for the task. In addition, some estimates include sandy lagoon areas with the coral reefs, while others exclude them. The scale of these problems is illustrated by two recent estimates of Indonesian coral reef area. One puts the total coral reef area in Indonesia at nearly 86,000

Diving on the reef, Bali.

km^2, while another puts it at only 42,000 km^2. Take your pick! For comparison, the areas of Java and New York State are both nearly 130,000 km^2, while those of Switzerland and the Netherlands are both about 41,000 km^2.

Fringing reefs

Reef scientists classify coral reefs into three main categories: fringing reefs, barrier reefs and atoll reefs. Many of the coral reefs visited by divers and snorkellers in Indonesia are fringing reefs. That is, they are reefs that hug the coasts of large, rocky islands (or in other countries, the coasts of the continental land mass). The popular reefs of north Bali, of Ambon and off Manado are all fringing reefs.

Barrier reefs

Islands can subside over geological timescales, and if this happens the fringing reefs will grow vertically upwards to keep level with the sea surface. As a result, a shallow lagoon will be formed between the reef and the sinking central island. A reef separated from its island in this way,

Sponges and soft corals as well as hard corals are important components of this reef at Sangalaki, East Kalimantan.

and forming a barrier between the open sea and the sheltered lagoon, is known as a barrier reef. A classic example of an island barrier reef is provided by the reef surrounding Pulau Besar in the mouth of Maumere Bay, Flores. An example of a shelf barrier reef (i.e. one associated with a subsiding section of a large land mass) is provided by the Spermonde Archipelago off Ujung Pandang.

The key to understanding the formation of barrier reefs from fringing reefs is to remember that reef-building corals contain symbiotic algae. Algae are plants and thus need sunlight to survive. Corals therefore thrive in shallow waters.

If the island they are growing around sinks, the corals will grow back up towards the surface and the light.

Atoll reefs

If an island with a barrier reef subsides so far that it disappears beneath the sea's surface altogether, then only the ring of coral reef will be left behind. This reef ring is an atoll. The central rocky island may continue to subside, even after it has disappeared below sea level. As long as the rate of subsidence is relatively slow, upward coral growth can continue, keeping the top of the atoll reef at sea level. Drilling by an oil company at Maratua (a raised atoll off East Kalimantan) revealed coral down to 429m below current sea level without reaching the underlying rock basement.

There are 55 atolls in Indonesia, out of a world total of about 425. The largest atoll in Indonesia is the massive Take Bone Rate (formerly Tijger Atoll) in the Flores Sea, south of Sulawesi. It has a total area of nearly 3,000 km^2 and is the third or fourth largest atoll in the world, depending on who is doing the measuring!

Darwin and Wallace

The simple classification of reefs into fringing reefs, barrier reefs and atolls, and the way in which one derives from the other by island subsidence, was first proposed by Charles Darwin. During the years 1832-36 Darwin sailed around the world as naturalist on HMS *Beagle*. While on that famous voyage, Darwin visited the Galapagos Islands and made the observations that led to his development of the theory of evolution by natural selection. It is less well known that during the same voyage Darwin made the geological observations that led to the modern theory of atoll formation.

The first person to apply Darwin's reef classification to Indonesian coral reefs was Alfred Russel Wallace, another great 19th century naturalist. He travelled throughout the 'Malay Archipelago' from 1854-62, visiting Watubela Island (formerly Matabello) in the Kei Islands in April 1860. Here he saw "the first example I had met with of a true barrier reef due to sub-

*Two latticed butterflyfishes, **Chaetodon rafflesi**, on a coral reef off Manado, north Sulawesi. These fish feed on small reef invertebrates as well as the polyps of corals themselves.*

sidence, as has been so clearly shown by Mr. Darwin." It is often forgotten that Wallace also developed the theory of evolution by natural selection completely independently from Charles Darwin, while travelling in the East Indies. However, Wallace is remembered for noting the differences between the essentially Asian land animals of western Indonesia and the Australasian animals of eastern Indonesia. He first proposed the boundary line between the two biogeographic regions that became known as Wallace's Line. Modern scientists recognize that there is an area of overlap, rather than a sharp boundary between the two regions, and sometimes call this area Wallacea. It encompasses Sulawesi, Bali and parts of Nusa Tenggara, and is the very centre of coral reef biodiversity. While the diversity of land animals in Wallacea is partly explained by the overlap of Asian and Australasian faunas, the great diversity of marine animals is partly explained by the overlap of Indian Ocean and Pacific Ocean faunas (pages 11-13).

Darwin's theory of atoll formation explains what happens to reefs when islands subside, but sometimes the opposite happens and islands rise. This is not uncommon in Indonesia, which is geologically very active. When an island is forced upwards, any corals lifted out of the sea will die, leaving the coral reef high and dry as an exposed limestone ridge. Kakaban and Maratua off East Kalimantan have already been mentioned as raised atolls. Rocky islands with fringing reefs that are raised are left with limestone terraces around their

sides. Such raised reef terraces can be found throughout Indonesia, from Sumatra to Irian Jaya, but they are commonest on the islands of Nusa Tenggara and the Banda Arc, as a result of active tectonic movements. One of the best examples is Kisar Island in the southern Laut Banda which has no fewer than seven raised reef terraces, the highest 150 metres above present sea level. While in the Kei Islands in April 1860, Wallace was, again, the first to recognize raised reef terraces in Indonesia, this time on Gorong Island.

Think pink

Although corals are the most important builders of coral reefs, they are certainly not the only ones. Other key components of coral reefs are the coralline algae (see page 48). These are plants that produce calcium carbonate just as corals do. They look more like pink concrete than most people's idea of a seaweed. Coralline algae are particularly important in areas where the corals themselves do not thrive. One such area is in cave entrances and overhangs, where there is insufficient light for coral growth. Another is the reef crest, or surf zone, where the surge is too powerful for corals to survive.

Opposite top: The bright colours often associated with coral reefs are not those of hard corals but of other reef inhabitants, such as sponges (like this white vase sponge, **Amphimedon**) and soft corals. Hard corals are usually dull greeny-brown, reflecting the colour of their symbiotic zooxanthellae. Here there is a hard plate coral with a white rim, showing that it is growing outwards faster than the zooxanthellae.
Opposite bottom: Coral reefs with large patches of **Acropora** branching corals are often home to pairs of long-nosed filefishes, **Oxymonacanthus longirostris**, which feed on the corals' polyps.

Where surge is strong, pieces of coral that may be broken off are soon ground down to sand. Animals too contribute to sand production. Parrotfishes are the most notorious sand producers on coral reefs. They feed by scraping off the thin layers of algae and other living matter growing on rocky surfaces. In doing so they inevitably take in quantities of the underlying coral rock. This is ground down by powerful teeth in their throats, so that the organic matter can be more easily digested. The waste product is coral sand.

Sand produced from coral rock is almost pure white, so the beaches of offshore coral islands are also white. In contrast, some areas adjacent to major volcanic islands have black volcanic sand. A host of other sand colours testifies to their varied origins. Divers should always take time to inspect the sandy patches between coral outcrops, or in lagoons adjacent to coral reefs. Although at first glance sand patches may appear lifeless, they are usually well stocked with interesting creatures. Look out for garden eels, flatfish, tiny sand dragonets, sea cucumbers and sea slugs.

Below: The Napoleon eel, **Ophichthus bonaparti**, *lives buried in sand, often adjacent to coral reefs. It normally shows only its head, and so does not reveal the distinctive black saddles that mark its body.*
Opposite: *Sand-dwelling dragonets are usually well camouflaged and very difficult to spot. Often it is only their irregular hopping movements that give them away. The fingered dragonet,* **Dactylopus dactylopus**, *is no exception, but the male has a brightly coloured dorsal fin which it uses in courtship or when disturbed.*

Living Together

For me the most fascinating thing about Indonesian coral reefs is the enormous number of different species living in intimate associations with each other. Biologists use the term *symbiosis* (which simply means 'living together') to describe the whole range of close relationships between different species, beyond the normal predator-prey interactions. These relationships may be mutually beneficial (*mutualism*), one partner may benefit while the other is unaffected (*commensalism*), or one side may benefit to the detriment of the other (*parasitism*).

Sea anemones and anemone fishes

A classic example of coral reef mutualism is the relationship between the anemone fish and its giant anemone host. Sea anemones have stinging tentacles, which they use both for their own protection and to catch their prey. Most types of fish are stung if they come in contact with

The pink anemone fish, **Amphiprion perideraion**, *is normally found in association with the magnificent sea anemone,* **Heteractis magnifica**.

the anemones' tentacles, and so they take care to keep clear. However, the anemone fish has a special mucus coating that enables it to live with the anemone without being harmed. It is therefore able to dive into the anemone's tentacles for safety when threatened by a predatory fish. In return for harbouring the anemone fish, the sea anemone itself gains some protection: the anemone fish drive off any butterflyfishes that would nibble at the anemone if they could.

Gobies and shrimps

Another classic example of symbiosis is the association between gobies and their partner shrimps. The shrimps act as miniature bulldozers, excavating burrows in the sand which they share with the gobies. From time to time the shrimps come right out of the burrow to browse nearby. The shrimps are blind, so would soon fall prey to predatory fishes if the gobies were not keeping a sharp lookout for danger. The shrimps keep one antenna touching the goby, so that they can pick up any warning twitch and dive back into the burrow immediately.

The gobies benefit of course by gaining a well-maintained home.

Corals and zooxanthellae

A third example of symbiosis is the association between reef building corals and the tiny single celled plants known as *zooxanthellae* (pronounced *zoo-zan-thel-ee*). The living tissues of the coral animals (or polyps) are packed with these microscopic plants. Like other plants they carry out photosynthesis, using the energy of sunlight to produce sugar and oxygen. The zooxanthellae benefit from this relationship by having a safe home. The corals benefit by having access to the relatively large amounts of oxygen and sugar that 'leak' from the zooxanthellae. These allow them to grow much faster than they would be able to do without the zooxanthellae. Indeed so important is this relationship for the corals that they cannot survive without their plant symbionts. For example, when sea temperatures become too warm (as they do in some areas during El Niño events), the corals become stressed and expel their zooxanthellae. The corals are said to be 'bleached', because they lose the colouration imparted by the zooxan-

Below: Goby, **Amblyeleotris steinitzi**, *and burrowing shrimp,* **Alpheus**, *form a mutually-beneficial symbiotic partnership.*
Opposite top: Indonesia abounds in examples of multi-layered symbiosis. Here three different types of animal are all living on a single bubble coral, **Plerogyra sinuosa**: *the small olive discs are flatworms,* **Waminoa**; *there is a symbiotic shrimp,* **Vir philippinensis**; *and an orangutan crab,* **Achaeus japonicus**, *which is itself camouflaged with a living coat of red algae.*
Opposite bottom: The crinoid squat lobster, **Alloga-lathea elegans**, *lives only on featherstars.*

A large isopod sea louse parasite on a cardinalfish, Apogon.

relationships. Parasites abound, although many of them are internal and so not readily seen. Fish suffer from a variety of external parasites, including crustacean isopods (fish lice, see also page 152) and copepods (see page 41). To be a fish with a louse, but no hands to do anything about it, must be torture indeed!

The terrible burden that external parasites impose is the reason why fish spend so much time visiting cleaning stations (see cleaner wrasses on pages 125 and 155, and cleaner shrimps on pages 83 and 85). Cleaning is itself another example of symbiosis.

Look closely and you will find other partnerships in action. In many cases it is easy to see what advantage one side is getting, but not how the other is benefiting. In other cases, the nature of the relationship may change according to circumstances. It is therefore often best to stick to the term symbiosis when describing such relationships, as it does not imply any judgement about benefits or costs.

thellae and reveal the underlying white of their limestone skeletons. If sea temperatures do not drop quickly, so that the corals can recover their zooxanthellae (and their colour), they will die.

Apart from the reef building corals, several other reef animals contain symbiotic zooxanthellae. Indeed, any sessile marine animal that lives in shallow water exposed to sunlight and has greeny-brown coloured soft tissues is likely to contain zooxanthellae. Examples include leather corals, giant sea anemones and giant clams. The advantage that zooxanthellae give to their hosts should be clear not only from the abundance of these animals in shallow reef waters, but also from the use of the word 'giant'. Giant sea anemones and giant clams reach sizes so much bigger than their relatives thanks to their zooxanthellae.

Parasites

While mutually beneficial partnerships attract our admiration, there are also plenty of less pleasant

*Opposite top: Pearlfishes, **Carapus**, live inside other animals. This individual has chosen a sea cucumber (the peacock sea cucumber, **Bohadschia argus**) as its host, but others live in starfishes, oysters and sea squirts. At least some pearlfishes are parasites, feeding on the gonads of their hosts. Those that never venture outside have lost all pigmentation, but this pearlfish looks as though it regularly pops its head out for a breath of 'fresh air'!*
*Opposite bottom: Remoras or suckerfishes, **Echeneis naucrates**, are usually associated with large pelagic animals such as sharks, rays, turtles and whales. However, they will occasionally attach themselves to reeffishes such as this blue-faced angelfish, **Pomacanthus xanthometopon**.*

Poisons and Mimicry

To protect themselves from the unwanted attention of predators, or to catch their own prey more efficiently, many marine animals are poisonous or venomous. The terms *poison* and *venom* are often used interchangeably, but they are different. Poisons are substances which cause pain, sickness or death if eaten. Venoms have these effects if injected.

As an example, pufferfishes are poisonous, because their skin and internal organs contain an extremely powerful poison, tetrodotoxin. This is the stuff that African and Caribbean witch doctors are said to use to turn people into zombies. Ironically, pufferfish flesh is safe to eat, and is considered to be a delicacy in Japan, where it is known as *fugu*. It can only be prepared for sale there by trained and licenced chefs, but accidents do happen, and deaths from *fugu* poisoning are not unknown.

Venoms

Sea snakes are not poisonous. They are venomous, because they inject

*The stonefish, **Synanceia verrucosa**, is perhaps the most venomous of all reef fishes. A sting from its dorsal spines is unbearably agonizing and can cause death.*

their toxins, using their fangs. Some sea snakes (page 156) are extremely venomous, their venom being ten times more potent than a cobra's. Why are sea snakes so venomous? After all, they do not need to kill several elephants with a single bite! One reason is that some sea snakes prey on moray eels, which are particularly tough customers and have over millions of years evolved great resistance to snake venom. Extremely potent venom is needed to kill these morays. Another reason for using such powerful venom is that it kills prey animals so quickly they have little chance to struggle and hurt the snake. Fortunately for divers and snorkellers, people are not on the snakes' menu. Sea snakes may sometimes be very inquisitive, but they are rarely if ever aggressive with swimmers, and in any case tend to have very small mouths and short teeth, which prevent them from biting large objects.

Other more commonly encountered venomous reef inhabitants include the catfish, and the scorpionfishes and their relatives (stonefishes and lionfishes). These fishes all have

venomous spines on their backs (page 112) which they use for defence. While the stonefish is the most feared of all venomous fishes, most stings to divers are caused by lionfishes and are inflicted on underwater photographers. Lionfishes turn their venomous spines towards danger. When confronted with an overenthusiastic underwater photographer they naturally turn their backs to the camera. This does not make for a good photo, so the

photographer will often wave their hand to one side of the lionfish. The fish will appear to oblige by turning its face to the camera. In fact it is turning its spines towards the threatening hand. The photographer, concentrating on framing the perfect shot, is too often unaware of what the lionfish is really up to!

Luckily, first aid for lionfish, and other fish, stings is simple. Most fish venoms are proteins, and as such they are denatured by heat. An effective treatment for most fish stings is therefore immersion of the affected part in hot water. This should not be boiling, or it will cause severe burns. About 45°C is okay. Keep immersed in hot water until the pain subsides.

In contrast to most fishes which use their spines for defence, cone

Opposite top and bottom: There are several species of stinger scorpionfish, **Inimicus,** *in Indonesia. They spend their time in or on sand and rubble bottoms, hidden from predators and prey alike. When discovered and threatened they flash the insides of their pectoral fins, as a warning that they are dangerous.*

Below: *Coral reef animals employ a wide variety of means to deter would-be predators. This six-striped soapfish,* **Grammistes sexlineatus,** *has a bitter tasting toxic skin mucus, which repels any predatory fish that does try to take a bite.*

shells (page 71) have a venomous barb which they use offensively to attack and immobilize their prey. Cones that feed on fish have particularly powerful venom, which can be deadly to humans. There are many species of cones, and it is difficult for the non-specialist to tell them apart. For this reason it is best to leave all cone shells untouched.

Warning: danger!

Having a potent poison or venom as a defence mechanism is all very well, but if a predator does not know you are well defended he may bite you anyway. Some animals therefore advertise their nasty natures with warning (or *aposematic*) colours. Such warning colours tend to be bright, and

contrasting bands are often present. For example, many sea snakes are banded, while poisonous flatworms (page 64) and sea slugs are brightly coloured and often striped.

Animals with warning colouration may still fall victim to an inexperienced predator that has not learnt to heed their warning. If two dangerous species have the same warning pattern, a predator has only to try one

Below: The magnificent pyjama sea slug, **Chromodoris magnifica**, *is advertising its obnoxious taste.*
Opposite top: All pufferfishes, including the diminutive sharpnose puffers like this **Canthigaster valentini**, *contain the deadly poison tetrodotoxin. They are avoided by most reef predators. This particular individual is afflicted with a parasitic copepod.*
Opposite bottom: The mimic filefish, **Paraluteres prionurus**, *pretends to be poisonous by mimicking the colouration of the sharpnose pufferfish,* **Canthigaster valentini**. *Note the subtle differences in patterning between the two, and the much broader dorsal fin in the filefish.*

and it will learn to avoid them both. For this reason, many dangerous animals have similar warning signs.

Mimicry

By way of contrast, some animals that are neither venomous nor poisonous have what appears to be warning colouration. They pretend to be dangerous, and take advantage of their false warning colours to avoid predation. This type of mimicry, where a harmless animal has similar colouration to that of a dangerous animal, is known as *Batesian mimicry*, after Henry Bates, the naturalist who first described it. Bates was a friend of Alfred Wallace (see page 25), and the two had collected natural history specimens together in Brazil before Wallace came to Indonesia.

Animals indulging in these types of mimicry tend to make a brave show of their warning signs in order to frighten off potential predators. However, for some animals the best means of avoiding predation is to blend in and avoid standing out. Many schooling fishes minimize their chances of being eaten by being just one in the crowd. Any blemish

Opposite top: It is not necessary to be dull to be camouflaged, as demonstrated by this splendid scorpionfish, **Scorpaenopsis**.
Opposite bottom: A pair of ghost pipefish, **Solenostomus cyanopterus**. *They are almost impossible to spot when living among sea grass. In slight currents they sometimes stand out because they do not drift with the flow like real bits of floating weed.*
Below: Sometimes an otherwise superbly camouflaged fish can be given away by the sharp outline of its eye. The crocodile fish, **Cymbacephalus beauforti**, *avoids this problem with its elaborate eye shades.*

or irregularity allows a predator to home in, so it is best to be exactly like all the other fish. Sometimes fishes of several species school together, and in these cases different species often have the same type of colouration. This is another type of mimicry, with different species of schooling animals having similar colouration in order to minimize predation, and is known as *social mimicry*.

Camouflage

Camouflage is the ultimate form of blending in, this time with the background. Bottom dwelling species, such as flatfishes and flatheads (pages 116 and 130) are all experts of camouflage. Many are able to adjust their colouration to blend in perfectly with the bottom they are lying on. It should be remembered that camouflage is useful not only for avoiding detection by potential predators, but also by potential prey. For example, frogfishes (page 106) use their superb camouflage to remain effectively hidden from view even

though they are sitting right out in the open. The use of camouflage (or more generally, looking like something harmless) as a means of avoiding detection by prey animals and thereby increasing the chances of catching them has been called *aggressive mimicry.*

Several varieties of crab make use of other organisms to disguise themselves. Sponge crabs of the family Dromiidae carry sponges on their backs for camouflage. The sponge is often much larger than the crab, which holds on to it with specially modified hind legs. Decorator crabs (of the spider crab family Majidae) pick up little bits of algae, sponge and other organisms and stick them on their shells for camouflage using special velcro-like projections. In addition to the examples shown on these pages, the orang-utan crab, *Achaeus japonicus*, which camouflages itself with a coat of red seaweed, is shown on page 33.

Some scorpionfishes have the best of both worlds. Overall they are superbly camouflaged. This allows them to avoid detection by both potential prey and potential predators. However, if they are spotted by a large predator (or disturbed by a diver!) they can flash their brightly coloured dorsal fins or inner sides of their pectoral fins, giving warning of their venomous spines (see page 38).

Left: Sponge crabs are most easily spotted at night, when they forage on the reef.
Opposite: This little decorator crab was photographed inside a barrel sponge.

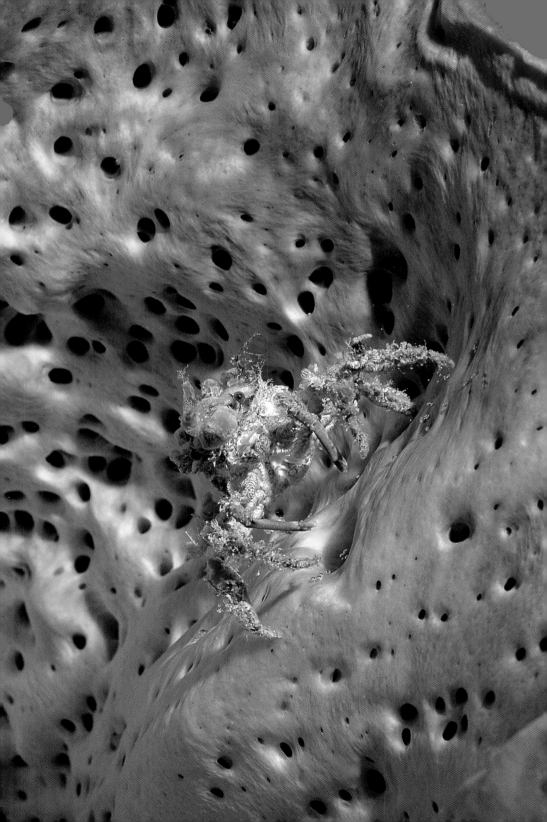

Marine Life

Marine Plants

1 ▲

2 ▲ 3 ▼

While plants are clearly the dominant organisms
mangrove forests and sea grass beds, they are ofte
inconspicuous on coral reefs. This is partly becau
many plants are microscopic, and tucked away insi
animals such as corals. Another reason is that many
the algae that do grow on the reef are immediate
grazed down by fish and other herbivores.

1 – *Encrusting coralline algae look more like pi*
concrete than plants. They thrive on coral reefs
high-energy or low-light environments which are t
harsh for corals to survive. This example, about 10 c
across, was photographed in a cave off Flores.

2 – *Plants use the energy of sunlight to produce foc*
and oxygen by photosynthesis. On calm sunny day
(when photosynthesis is in top gear) seaweed surfac
are festooned with oxygen bubbles.

3 – *Despite its unusual appearance and its name, th*
sailor's eyeball, **Valonia ventricosa**, *is a single*
celled green alga.

4 – *A mangrove seedling, or hypocotyl, takes root.*

4

Sponges

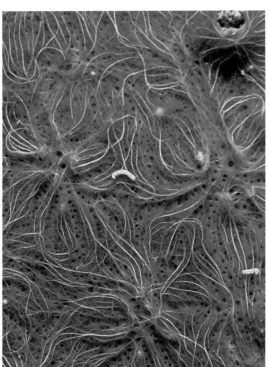

Despite their inanimate appearance, sponges are indee animals. They are filter feeders, picking out any edib particles from water drawn in through the many micr scopic holes that cover their surfaces. These entran ports are named ostia, *after Ostia Antica, the entry po for the city of ancient Rome. The water passes throug the body in a one-way system, passing back out throug the relatively small number of large holes (or oscul that in many species are clearly visible to the naked ey*

5 – *Sponges have skeletons composed of rods (spicules) of silica or calcium carbonate. In the case this encrusting sponge,* **Nara***, there are also high visible collagen fibres.*

6 – *Many sponges grow vertically. By sticking u into the current, water is drawn through the spong like smoke through a chimney.*

7 – *Barrel sponges,* **Xestospongia***, like this individu photographed off West Sumatra, are the giants of t sponge world. Large specimens may be crowded wit featherstars and other animals, making use of t sponge as a lofty perch.*

5 ▲ 6 ▼

7

Hard Corals

Hard corals are the backbone of the reefs. The specie[s] which contain symbiotic algae (zooxanthellae) ca[n] grow faster than their competitors, but because th[e] algae need sunlight they are limited to shallow water[s]. The greatest profusion of hard corals is therefore t[o] be found on shallow reef flats and the uppermost part[s] of reef slopes. Fast growing species often have a whit[e] rim: the greeny-brown zooxanthellae have not ha[d] time to move into the newly formed coral tissue.

8 – Many corals synchronize their spawning with th[e] phase of the moon in order to maximize the chances [of] fertilization and larval survival. This coral, **Echino-pora**, in Maumere Bay, Flores was photographe[d] spawning eight days after the full moon in Octobe[r.]

9 & 10 – The common orange coral that lives unde[r] overhangs on many reefs is often known simply a[s] **Tubastrea**. In fact there are several closely relate[d] species, of both **Tubastrea**, and **Dendrophyllia**. [It] is not easy to tell them apart. At night the polyps [of] **Tubastrea** corals expand, forming luxuriant orang[e] carpets (right). During the day, the polyps are retracte[d] exposing the underlying hard stony skeleton (below[.]

8 ▲

9 & 10 ▼

11 ▲

12 ▲ 13 ▼

Different coral species have different shapes, or growth forms. However, even within one species precise shape may vary according to local environmental conditions.

11 – *A massive coral,* **Lobophyllia**. *When referring to growth forms, the term massive implies a hemispherical shaped coral, not necessarily a large coral.*

12 – *A robust branching coral,* **Pocillopora**.

13 – *The commonest branching corals are those of the genus* **Acropora**. *They have characteristic large polyps at the end of each branch which are a specialized adaptation for rapid growth.*

14 – *The leaf-like, or foliaceous, growth form of the coral,* **Turbinaria reniformis**, *allows it to make maximum use of incoming sunlight.*

14 ▶

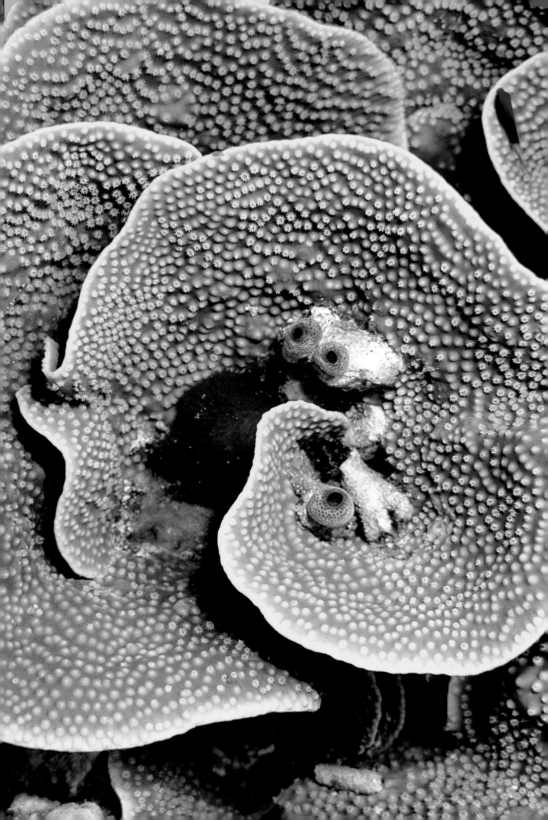

15 ▲

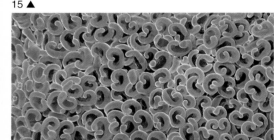

16 ▲ 17 ▼

15 – *The distinctive coral,* **Symphyllia**, *is abundan[t] on Indonesian reefs.*

16 – *The polyps of the anchor coral,* **Euphylli[a] ancora**, *have tentacles with characteristic anchor- o[r] kidney-shaped tips.*

17 – *The long white-tipped tentacles of* **Euphylli[a] glabrescens** *make it look superficially like a se[a] anemone. However, a hard coral skeleton lie[s] underneath.*

18 – *A close-up photograph of a massive cora[l]* **Acanthastrea**. *Each little ring is a corallite, the hom[e] of a single coral polyp. Identifying many species [of] corals from underwater photographs is problemati[c] because it requires information on the structure of th[e] underlying skeleton, which is obscured by livin[g] tissues.*

18 ▶

Soft Corals

*The soft corals, as the name suggests, lack the har
stony skeletons of their reef-building relatives
Instead, they have tiny rigid rods (or* sclerites) i
*their tissues, and in some species use hydrostati
pressure to keep themselves erect. Soft corals are als
known as octocorals, because each individual poly,
animal has eight tentacles (unlike hard coral polyp
which have multiples of six).*

19 – Seapens, such as this **Virgularia**, *are relative
of soft corals. They live on sandy and muddy bottoms
and use the retractable feathery 'quill' (or rhachis) fo
filter feeding.*

*20 – Several types of soft coral have a dull greeny
brown colouration. These varieties, including thi
soft coral* **Nephthea**, *tend to be found in shallow ree
areas that are well exposed to sunlight.*

*21 – Many reefs are dominated by dull coloured har
corals. It is soft corals that provide the spark of colour
There are dozens of different species on Indonesia
reefs, most of which cannot be identified to specie
from photographs. This is a type of* **Dendronephthya**

19 ▲ 20 ▼   21 ▶

22 ▲

23 ▲ 24 ▼

Sea Anemones

Sea anemones have multiples of six tentacles, s despite superficial appearances, are more closel related to the hard corals than the soft corals. Th giant sea anemones have symbiotic algae in thei tissues, as well as anemonefish partners.

22 – *The anemone hermit crab,* **Dardanus pedun culatus**, *carries a number of anemones,* **Calliactis** *on its shell. The anemones benefit from this relationshi, by being carried to new feeding areas, while the her mit crab gains the protection of the stinging anemones*

23 – *The stinging anemone,* **Actinodendron**, *pack a powerful sting. Thus many symbiotic shrimps an crabs can often be found sheltering among its arms*

24 – *Haddon's sea anemone,* **Stichodactyla had doni**, *is found in lagoons and other sandy areas. Thi individual is playing host to some juvenile three spo damselfishes,* **Dascyllus trimaculatus**.

25 – *The distinctive tentacles of the bulb-tentacle see anemone,* **Entacmaea quadricolor**. *It is host to a least six different species of anemonefish in Indonesia*

25 ▶

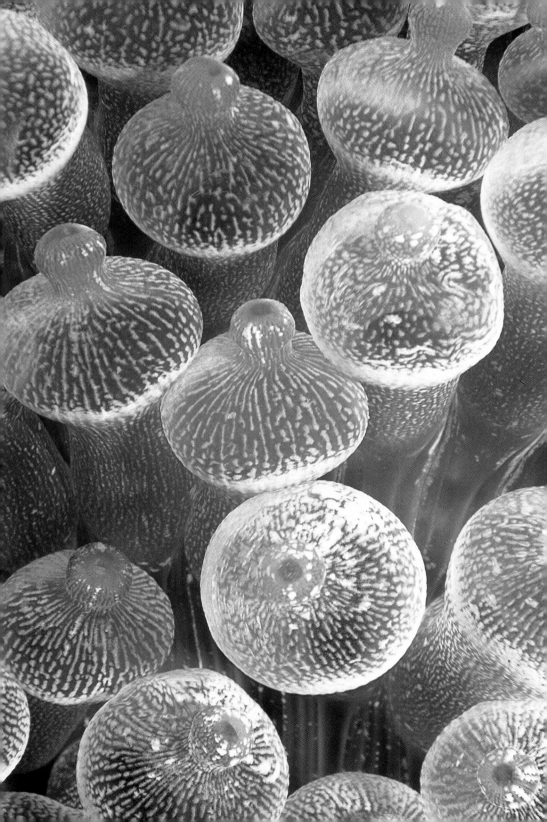

Sea Fans and Black Corals

Both sea fans (or gorgonians) and black corals hav[e] internal skeletons made of highly flexible, horn-lik[e] materials, which allow them to bend with the curren[t] and thus avoid damage. Despite these similarities th[e] two types of animal are not very closely related. Se[a] fans are octocorals, related to soft corals, while blac[k] corals are hexacorals, related to hard corals.

*26 – A close-up photograph of a sea fan, **Acaly[...] cigorgia**, shows many white polyps extended. Eac[h] polyp has the eight tentacles typical of octocorals.*

*27 – Sea fans, such as this relatively small **Sube[r] gorgia mollis**, grow out from the reef, perpendicul[ar] to the prevailing current. Therefore nearly all a[re] oriented with their long axis vertically. Divers shou[ld] take care in those few areas where sea fans are oriente[d] with their long axis horizontally, since this indicate[s] that there may be potentially dangerous down current[s.]*

*28 – It comes as a surprise to many divers to learn th[at] black corals, **Antipathes**, are brown! The semi[-] precious black material is the polished skeleton [of] bushes such as this.*

26 ▲ 27 ▼

28

62

Flatworms

The flamboyant flatworms, with their gaudy colour and frilled edges, are miniature stars of the reef. Their showy appearance warns that they are poisonous or bad tasting. Flatworms are simple animals. Because they are flat they can absorb all the oxygen they need through their body surface, so they lack both gills and a blood circulatory system.

29 – The exquisitely patterned flatworm **Pseudoceros bedfordi** is widely distributed in the Indo-Pacific, but was first collected at Singapore.

30 – Flatworms come in a wide variety of outrageous colours. This undescribed species of **Pseudoceros** has also been recorded from Papua New Guinea.

31 – This fabulous flatworm, **Pseudoceros ferrugineus**, is advertising the fact that it is not good to eat.

32 – This striped flatworm was photographed in Kakaban Lake, East Kalimantan. It looks similar to the well-known and widespread three-striped flatworm, **Pseudoceros tristriatus**, although it differs in some details and may be a new species.

29 ▲

30 ▲ 31 ▼

32

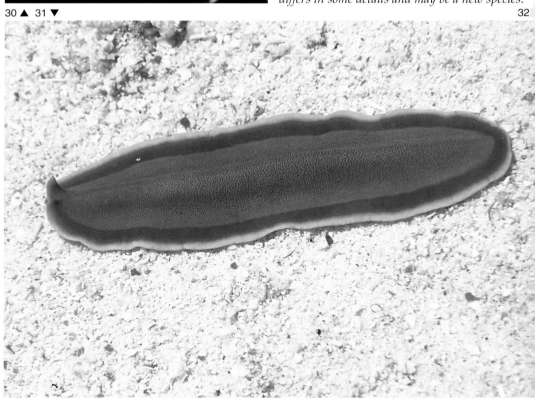

33 ▲

34 ▲ 35 ▼

Fan Worms

There are many different types of worm to be seen on Indonesian reefs, but the most familiar must be the feather-duster worms (also known as fan or Christmas tree worms). These do not look at all like most people's idea of a worm, for the worm-like body is hidden in a tube, with only the crown of feeding tentacles protruding.

33 – This unusual looking 'Winston Churchill' worm has a pair of bristly feeding tentacles. It was photographed near Derawan in East Kalimantan. It is a sabellarid worm, but the species is not known.

34 – The identification of reef worms is still at an early stage, and it is not possible to put a name to many species photographed with any degree of certainty. This large fan worm (about 8 cm across) may be a species of **Sabellastarte**.

35 – Every underwater photographer's macro favourite, the christmas tree worm, **Spirobranchus giganteus**.

36 – The fragile colonial fan worm, **Filogranella elatensis**, is found in sheltered locations on the reef slope.

36

Sea Snails

The diversity of sea snails in Indonesian seas is quite astonishing. There are thousands of species. Many are small, and remain hidden in the reef or under sand and mud. All are eaten by a variety of predators, so even many of the most familiar types tend to stay hidden by day, only emerging to feed at night.

37 – This volute snail, **Cymbiola vespertilio**, is active by night. It is carnivorous, preying on other snails and invertebrates which it envelopes with its foot.

38 – The extended foot and mantle of this false cowry, **Calpurnus verrucosus**, cover the shell and help it to merge in with its leather coral food and habitat. This species was first described scientifically in 1705 by the Dutch naturalist Rumphius, from specimens collected in Ambon.

39 – This splendid individual is a map cowry, **Cypraea mappa**. It is only found in the western Pacific.

40 – The tiny pink-spotted allied cowry, **Primovula**, is less than 1 cm long. It lives exclusively on soft corals of the genus **Dendronephthya**.

37 ▲

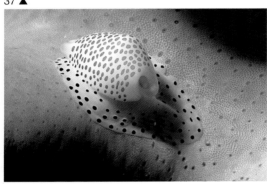

38 ▲ 39 ▼

40

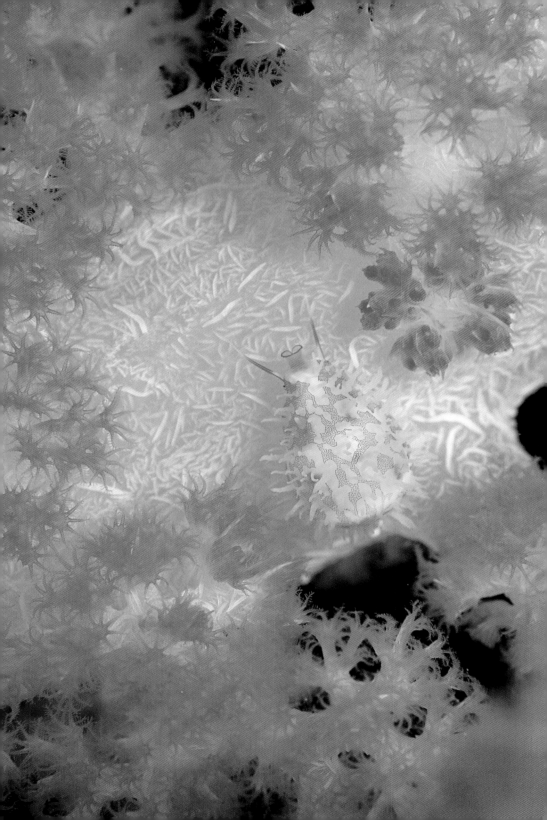

41 – *The worm snail,* **Serpulorbis grandis**, *i*
cemented 'upside-down' to the reef. With its foo
pointing upwards, it secretes a mucus net with whic
it traps tiny food particles.

42 – *The murex shells, like this* **Chicoreus**, *ar*
predators of other snails, bivalves and variou
invertebrates. Murex snails from the Mediterranea
were the source of Tyrian purple, the colour favoure
by Roman emperors and other ancient royalty.

43 – *Stromb shells are characterized by the 'strom*
notch', through which the right eye protrudes. Thes
Strombus, *like other related species, are algal grazer*

44 – *Cone shells have a venomous dart which they us*
to immobilize their prey. Some species, including th
textile cone, **Conus textile**, *can kill humans.*

41 ▲ 42 ▼

43 ▲ 44 ▼

Sea Slugs

Sea slugs are popular with many divers and especiall underwater photographers. As with flatworms, thei gaudy colouration advertises the fact that they ar poisonous or have a noxious taste. Most are ver small, often less than 2 cm in adult size. They are ofter known by the scientific name of **nudibranchs** (whic means naked gills, i.e. the gills are exposed, not tucke away underneath the body or in a shell).

45 – The egg ribbon of a Spanish dancer, **Hexabranchu sanguineus**. The bright colour warns fishes that thes eggs should be avoided.

46 – The Spanish dancer, **Hexabranchus sanguineus** is most likely to be seen at night. It swims in a styl reminiscent of a flamenco dancer.

47 – This beautiful western Pacific sea slug, **Phyllidi willani**, was named in honour of nudibranch exper Richard Willan in 1993.

48 – The mouse sea slug, **Kentrodoris rubescens** is usually found in relatively shallow water. Thi individual was photographed in 22m.

45 ▲ 46 ▼

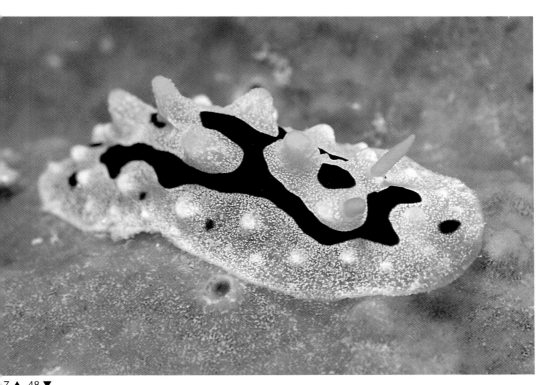

49 – *The beautiful sea slug* **Hypselodoris bulloc** *is a familiar sight on Indonesian reefs. It shows considerable colour variation, and there may in fact several similar species masquerading under this o name.*

50 – *The brick red sea slug* **Chromodoris reticula** *has a simple but striking colour pattern.*

51 – *The common western Pacific sea slug* **Chrom doris coi** *waves the front of its mantle skirt up an down as it moves, exposing the purple underside.*

52 – *A pair of sea slugs,* **Chromodoris lochi**, *feedin on a brown sponge (perhaps* **Callyspongia**). *S slugs tend to have very specific diets, with differer species specializing in different prey. To cope wit their particular diets, each species has a uniqu arrangement of teeth, which scientists find ver useful in identification.*

49 ▲ 50 ▼

Oysters and Clams

*These relatives of the sea snails and sea slugs a. called **bivalves** because their shells are made in tu parts, or valves. Many are edible, the favourite pa being the muscle that pulls the two halves of the she together.*

53 – *Zig-zag oysters, **Lopha**, are usually coated l encrusting sponges.*

54 – *The reef clam, **Spondylus varius**, is a famili inhabitant of reef caves. These bivalves are ofte adorned with sea squirts, sponges and other creature*

55 – *This reef oyster, **Hyotissa**, has been photographe while spawning. Sperm are being expelled from on side. To maximize the chances of fertilization, mo individuals of one species will spawn at exactly tl same time, perhaps only once a year.*

56 – *Giant clams like this **Tridacna** have their sof fleshy mantles exposed to the sunlight. This allou the symbiotic algae (zooxanthellae) inside to carr out photosynthesis. The clam benefits with a share the food produced by its guests.*

53 ▲

54 ▲ 55 ▼

56 ►

Cuttlefish and Octopus

Cuttlefish, octopus and also squid belong to the group known as cephalopods (or head-foots) because the legs grow out from the head region. The octopus has eight legs, while the others have ten. Intelligent predators, they have well developed eyes and sophisticated behaviour.

57 – *Looking like a creature from another planet, this little cuttlefish, **Sepia**, struck a defiant defence pose when confronted by an underwater photographer several hundred times its size.*

58 – *The skeleton of the cuttlefish, **Sepia latimanus**, is the well-known cuttlebone. The calcium carbonate 'bone' is permeated with air cavities, which allow the cuttlefish to maintain neutral buoyancy.*

59 – *The common reef octopus, **Octopus cyanea**, can change its colour to match its surroundings. It has no internal skeleton at all, and so is able to squeeze through tiny reef crevices to escape predators or to chase prey.*

60 – *This ornate banded octopus, **Octopus** sp., is one of a group of species that have not yet been named scientifically. It was photographed at Tulamben, Bali.*

57 ▲ 58 ▼

59 ▲   60 ▼

Crabs

Crabs are high on the menu of many reef predators including fishes and octopus. Perhaps for this reason many species of tropical reef crab contain deadly poisons. Therefore, none should be eaten without local advice. For the same reason, many remain well out of sight during the day, and only emerge under cover of darkness.

61 – The large spotted reef crab, **Carpilius maculatus**, hides inside the reef by day, only emerging at night.

62 – Those crabs that are out in the daytime are usually associated with other animals that can provide them with protection. This commensal crab **Lissocarcinus** is safe among the tentacles of its host tube anemone.

63 – The gorgonian crab, **Xenocarcinus conicus**, is found on sea fans and black corals. Its colour change to match that of its background. The yellowish markings on this crab's legs mimic the polyps of the sea fan.

64 – The seapen crab, **Porcellanella triloba**, is found in pairs on seapens (page 58). One pair of feeding appendages are modified into strainers, which the crab use to pluck plankton from the water.

61 ▲

62 ▲ 63 ▼

64 ▶

Shrimps

Shrimps are eaten by fish as well as people, so have evolved many strategies to keep themselves out of trouble. Some are nocturnal, some stay permanently hidden, some live in association with stinging anemones, while others are almost completely transparent and therefore virtually invisible. Cleaner shrimps, on the other hand, make themselves indispensable and highly visible to reef fishes

65 ▲

65 – *Hingebeak shrimps,* **Rhynchocinetes***, are abundant on Indonesian reefs, but normally only venture out at night. As the name suggests they have an enlarged head spine (or beak) that can be folded down*

66 ▲ 67 ▼

66 – *The head of this shrimp,* **Periclimenes kororensis***, is white to match the tips of the tentacles of its host mushroom coral,* **Heliofungia actiniformis***.*

67 – *The imperial shrimp,* **Periclimenes imperator***, lives symbiotically on other reef invertebrates. This one was photographed on a sea cucumber (page 98) but it is also found on sea slugs and large starfish.*

68 – *The Ambon cleaner shrimp,* **Lysmata amboinensis***, attracts passing fishes with its long white antennae*

68 ▶

82

69 – *There are two main types of mantis shrimp encountered by divers and snorkellers on coral reefs: spearers and clubbers. This burrow-dwelling mantis,* **Lysiosquilla**, *is a spearer. One of its many pairs of limbs is modified into vicious comb-like rakers, with which it can snatch an unwary fish out of the water above in just a fraction of a second.*

70 – *The common reef mantis,* **Odontodactylus scyallarus**, *is a clubber. It has a modified pair of limbs which it uses to beat its prey into submission and also to attack competitors of the same species. These shrimps have a nasty temper to match their fighting abilities, so they should not be handled!*

71 – *A pair of tiny shrimps,* **Allopontonia iani**, *are home among the venomous spines of the fire urchin* **Asthenosoma varium**. *At least one other species of shrimp and a snail live symbiotically with this sea urchin, taking advantage of the protection it affords from potential predators.*

72 – *This beautiful transparent cleaner shrimp is* **Urocaridella antonbruunii**. *It is found in reef caves and crevices, sometimes in large groups, where it waits in mid-water for fish to visit and be cleaned.*

69 ▲

70 ▲ 71 ▼

72 ▶

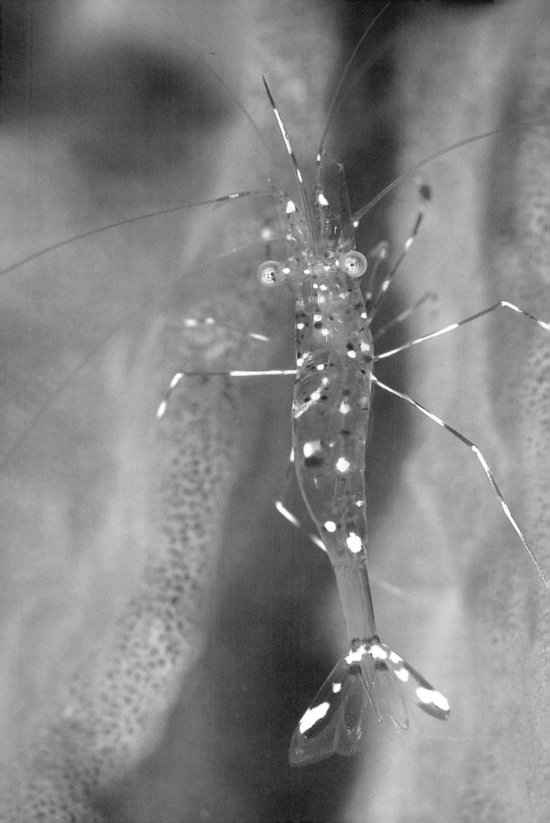

Lobsters

Eating lobster is a real treat for most people. There i[s] considerable demand for them not only within Indonesia itself (where lobster is known as udang karang) but also from export markets such as Japan and the United States.

*73 – The giant slipper lobster, **Scyllarides haanii** has a heavily armoured appearance, but is relatively docile and can be easily approached as it lumbers around the reef at night like a miniature tank.*

*74 – The common spiny lobster, **Panulirus versicolor** is widely caught for its meat. There are several species of spiny lobster in Indonesia, but this is the one most often seen on coral reefs. By day it hides in caves and crevices, with only its long antennae showing.*

*75 – The slipper lobster, **Parribacus antarcticus**, i[s] most likely to be found on cave roofs at night. It i[s] found in all tropical waters, so the specific nam[e] 'antarcticus' is particularly inappropriate.*

73 ▲   74 ▼   75

76 ▲

77 ▲ 78 ▼

Starfishes

Starfishes or seastars are for many the quintessential symbols of the sea. They, and their relatives, typically have five-sided symmetry, which is seen most clearly in the five-armed starfishes. They are also characterised by a spiny skin, from which the entire group (starfishes, brittlestars, featherstars, sea urchins and sea cucumbers) gets the name Echinoderm.

76 – *The cushion star,* **Culcita novaeguineae**, *may not look like a starfish at first glance. But closer inspection reveals that it does indeed have five sides. The tiny symbiotic shrimp,* **Periclimenes soror**, *lives on its under side. If you pick up a cushion star to look underneath, do so gently to avoid damaging its delicate tube feet.*

77 – *The thick armed starfish,* **Choriaster granulatus**, *is sometimes host to parasitic pearlfishes (page 35).*

78 – *The long-armed starfish,* **Iconaster longimanus**, *is one of the most beautiful Indonesian starfishes. Relatively uncommon, it may be found on deeper reef slopes.*

79 – *The blue reef starfish,* **Linckia laevigata**, *is one of the commonest reef starfishes in the western Pacific.*

79 ▶

80 – *The giant starfish,* **Thromidia catalai**, *grows t[...] at least 40 cm in diameter. It is found on coral-rich re[...] drop-offs. The tips of the arms appear green under water, only revealing their red colouration whe[...] brought to the surface or illuminated with artifici[...] light. This is due to the presence of pigments simila[...] to those in our blood (which also appears gree[...] underwater).*

81 – *This distinctive starfish,* **Echinaster callosus** *is widely distributed in the Indian and western Pacifi[...] Oceans.*

80 ▲

82 – *The surface of this particular species of starfis[...]* **Echinaster luzonicus**, *is often infested with sym[...] biotic benthic comb jellies,* **Coeloplana astericola** *Although these may look like flatworms (page 64[...] they are actually distant relatives of jellyfish. The[...] feed by shooting out stinging tentacles to entra[...] plankton.*

83 – *This starfish,* **Fromia monilis**, *is common o[...] coral reefs throughout Indonesia. There are severa[...] similar-looking species, which can make identificatio[...] to species rather tricky.*

81 ▲   82 ▼   83

Brittlestars

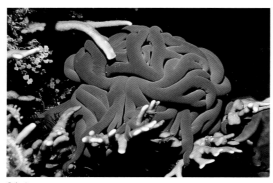

84 ▲

85 ▲ 86 ▼

Relatives of the starfishes, brittlestars can be distinguished by their spindly, highly flexible arms. Despi[te] their spiny and unappetizing appearance, brittlestar[s] are preyed upon by many species of fish, so most re[ef] varieties stay well hidden under rocks and in crevice[s].

84 & 85 – *The spectacular basket stars,* **Astroboa**, a[re] *specialized, nocturnal filter-feeding brittlestars. B[y] day, they hide within the reef, with arms well wrappe[d] up (top). After sunset they climb up on to a prominen[t] part of the reef and unfold their arms to feed (centre[). They are highly sensitive to light, and at the first flas[h] of a night diver's torch will start to fold up.*

86 – *This brittlestar,* **Ophiothrix**, *shows the typic[al] spindly arms which are clearly differentiated fro[m] the central disc-like body. The numerous spines offe[r] some protection against predatory fishes.*

87 – *Brittlestars that are visible by day are often woun[d] tightly around sedentary reef invertebrates such a[s] sponges, or, as in this case, soft corals, to avoid th[e] attentions of predatory fish. At night they loosen one o[r] more arms to feed on the rich nocturnal plankton.*

87 ▶

Featherstars

Feather stars, or **crinoids**, *have two sets of legs. There is a set of small 'feet' underneath which are only used for grasping, and the second set of large feathered legs which are used for both filter feeding and walking. Several species of feather star are nocturnal.*

88 – *Featherstar arms typically branch off at the base from five primary arms, so normally occur in multiples of five. This* **Stephanometra** *has 25 arms.*

89 – *Featherstar arms bear numerous side branches or* pinnules, *which in turn carry dozens of tiny hair like projections. Between them they form an almost impenetrable net for their tiny zooplanktonic food.*

90 – *Different species of featherstar are often distinguished by the minute details of their segments and spines, so they can be impossible to identify from photographs. However, both the colour and form of this blood red featherstar,* **Himerometra robustipinna**, *are distinctive.*

91 – *The featherstar* **Oxycomanthus bennetti** *is a common and most conspicuous coral reef species.*

88 ▲

89 ▲ 90 ▼

91

Sea Urchins

Sea urchins are grazers, feeding on both algae and encrusting animals with the powerful jaws that protrude from their undersides. The urchin's shell fills up with gonads before breeding, making it an appetizing snack for any fish that can penetrate its spiny defence. Large wrasses are surprisingly adept at eating them, so many urchins hide in reef crevices during the day, and only come out after sunset, when the wrasses are asleep.

92 – This long spined urchin, **Diadema savignyi**, is host to a symbiotic needle shrimp, **Stegopontonia commensalis**.

93 – Sand dollars, **Clypeaster**, spend much of their time buried beneath the sand. The five-sided symmetry of echinoderms is partly lost in these species, but remains in the pattern of grooves on the upper surface.

94 – This thick spined imperial sea urchin, **Phyllacanthus imperialis**, has lost a number of its original spines, and is now regenerating replacements.

95 – The sea urchin **Tripneustes gratilla** is found in shallow bays and lagoons, especially near sea grass beds.

92 ▲ 93 ▼

4 ▲ 95 ▼

Sea Cucumbers

96 ▲

97 ▲ 98 ▼

Sea cucumbers, or **holothurians**, *are important mem bers of sand and muddy bottom communities. They ingest large quantities of sand and mud, digest out the organic matter and pass out the remains, leaving a tell tale trail of faeces. Several species are prized in Chinese cuisine as bêche-de-mer, or trepang. Sea cucumbers are relatives of the starfishes, although their five-sided symmetry is only apparent in cross section.*

96 – The large burrowing sea cucumber, **Neo thyonidium magnum**, *lives buried in the sand, with just its crown of long feeding tentacles protruding. The tentacles trap plankton, which is licked off as each tentacle in turn is folded down into the central mouth*

97 – The beautiful red-lined sea cucumber, **Thelenota rubrolineata**, *is a large and distinctive species.*

98 – Found on exposed rocky coasts, **Actinopyga mauritiana** *has a particularly powerful grip.*

99 – A close-up photograph of the skin of the peacock sea cucumber, **Bohadschia argus**, *showing the pattern of 'eyes' which give it its name.*

99

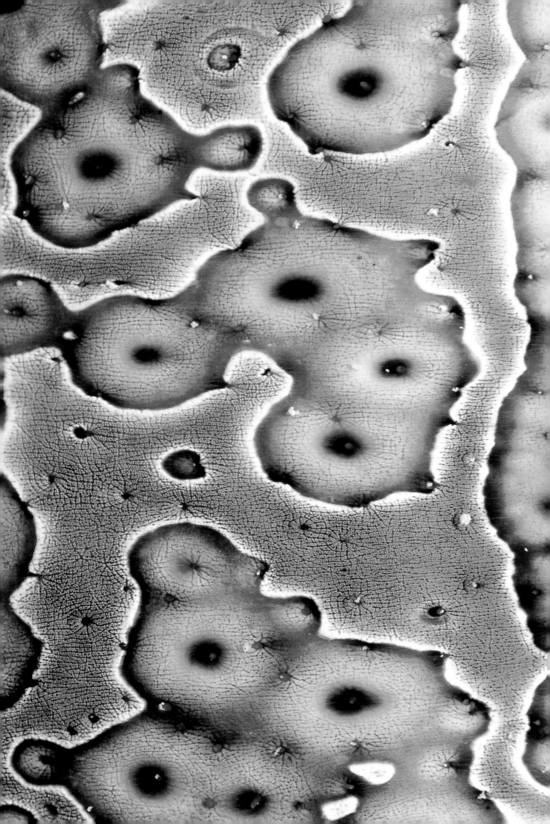

Sea squirts

Sea squirts, or **ascidians**, bear a superficial resemblance to sponges. Much of this is due to the fact that both are sedentary filter feeders. However, from a biological point of view, sea squirts are very much more advanced animals than sponges. Some sea squirts are solitary, some grow together in groups, and some are fused to form colonies.

100 – The brightly coloured solitary sea squirt **Polycarpa aurata** is a familiar sight on Indonesian reefs.

101 – Small sedentary reef animals sometimes grow together in what renowned divers Valerie and Ron Taylor have called sea posies. They develop in areas with strong water flow, and the Taylors rate Komodo as the best place in the world to find them. This delightful posy contains at least three species of sea squirts: the pale blue **Clavelina moluccensis**, together with one yellow and one bright red colonial didemnid species.

102 – These splendid blue sea squirts, **Rhopalaea crassa**, show the two openings, or siphons, typical of these animals. Water is drawn in through the top siphon, filtered internally, and expelled through the side siphon.

100 ▲ 101 ▼

102 ▶

103 ▲

Sharks and Rays

Sharks, and their relatives the rays, are only distantly related to other fishes. They differ in a number of characteristics, having a skeleton made of cartilage instead of bone, having separate openings for each gill instead of a bony flap covering all the gills, and producing small numbers of well-developed young. This method of reproduction, combined with slow growth, means that shark and ray populations are very easily overfished.

103 – *The blue-spotted fantail ray,* **Taeniura lymma**, *is relatively common on Indonesian reefs.*

104 – *A spotted maskray,* **Dasyatis kuhlii**, *hides itself in the sand. Many stingrays feed on clams and other invertebrates buried in the sand and mud.*

105 – *This school of scalloped hammerhead sharks,* **Sphyrna lewini**, *was photographed in 40 metres near Kakaban, off East Kalimantan.*

106 – *A giant manta ray,* **Manta birostris**, *feeds on the tiny planktonic animals that it filters from the water. Note the little white manta remora,* **Remorina albescens**, *near the manta's eye.*

106 ▶

104 ▲   105 ▼

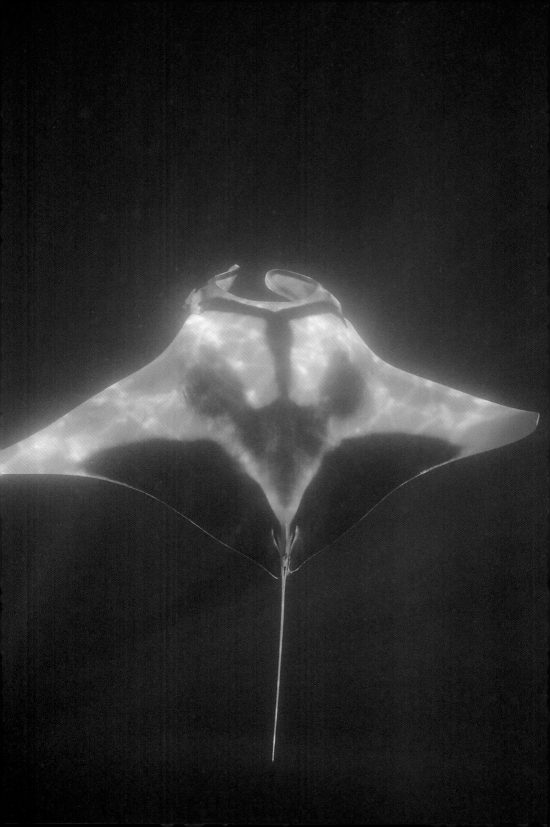

Eels

Moray eels spend much of their time hidden in reef crevices, where they feed on small fishes and other animals. Several species rest with their heads protruding from their lairs, often with one or two cleaner shrimps in attendance. Some species are more active by night, when they are more likely to be seen out and about on the reef.

107 & 108 – *The exquisite blue-ribbon or ghost moray,* **Rhinomuraena quaesita***, changes colours as it develops. Juveniles are mainly black, males are bright blue and yellow (left), while mature females are yellow (below).*

109 – *The starry moray,* **Echidna nebulosa***, was named scientifically in 1789, from an Indonesian specimen. It is a small, shallow-water species which feeds mainly on crabs.*

110 – *Garden eels live in sandy areas exposed to the currents. Keeping their tails anchored in their burrows, they feed on the plankton which drifts past. This particular species,* **Heteroconger hassi***, is named in honour of the famous diving pioneer Hans Hass.*

107 ▲ 108 ▼

09 ▲ 110 ▼

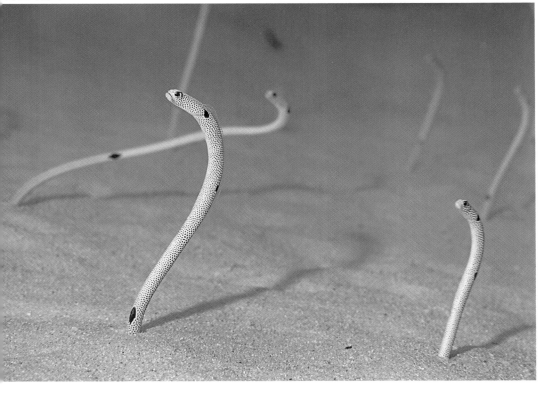

Frogfishes

111 ▲

Frogfishes are masters of camouflage, and are usuall extremely difficult to spot. To find one is therefore real challenge. Sometimes, on lucky rare occasions, a individual of a particular colour will position itse against a contrasting background, making it relativel easy to see. About 20 species can be found in Indonesi

111 – Antennarius nummifer is usually known i English as the coin-bearing frogfish, because of th circular mark on its side.

112 – The warty frogfish, **Antennarius maculatu** is so called because its body is covered with numerou wart-like swellings.

113 – Frogfishes have a special fishing rod and lu attached to the tops of their heads, which they use t tempt small fishes within striking range. This painte frogfish, **Antennarius pictus**, has a lure like a shrimp

114 – Colour is not a good guide to the identificatio of frogfishes. This species, **Antennarius coccineu** is known as the scarlet frogfish because it is usuall red, but several colour variations do exist.

112 ▲ 113 ▼   114 ▶

Nocturnal Fishes

Many nocturnal fishes are red. Red light is rapidl[y] absorbed underwater, so red fishes appear black a[t] night and are effectively invisible. They also tend t[o] have large eyes, with retinas specially adapted t[o] make use of what little light is available.

115 – *The bigscale soldierfish,* **Myripristis berndt[i]** *Soldierfishes feed on the plentiful nocturnal planktor[.]*

116 – *It is always worth taking a look for unusual crea[-] tures in very shallow water, particularly under jettie[s] This bigeye,* **Heteropriacanthus cruentatus**, wa[s] photographed under a jetty near Padang, West Sumatr[a]*

117 – *These blue-eyed cardinalfishes,* **Apogo[n] compressus**, *pass the day sheltering among the spine[s] of sea urchins. Many Indonesian cardinalfishes ar[e] not red, but several species are the colour of a Roma[n] Catholic cardinal's robes. It was for this that the famil[y] as a whole was named.*

118 – *A sabre squirrelfish,* **Sargocentron spiniferum** *Of the 16 or so species of squirrelfish found in Indonesi[a] this is the largest. The cheek spines are mildly venomou[s]*

Pipefishes

The pipefishes, and their relatives the seahorses, have elongated bodies encased in bony rings. These fish have a unique breeding system. The males brood the eggs until they hatch, holding them in a special belly patch or pouch.

119 – These fishes are known as sea horses in Indonesia ('kuda laut' or 'ikan kuda') as well as in English. When this particular species was first given a scientific name by Pieter Bleeker in 1852 he made use of part of the Indonesian name, calling it **Hippocampus kuda**.

120 – The ornate ghost pipefish, **Solenostomus paradoxus**, is fairly common in Indonesia, but very difficult to find. Look for it hiding among featherstars.

121 – This pipefish, **Corythoichthys schultzi**, is one of several similar-looking species that are relatively common on Indonesian coral reefs.

122 – The white pipefish **Siokunichthys nigrolineatus** is only found living in association with solitary mushroom corals, **Fungia**. What benefit the coral gets from this symbiotic relationship is not clear.

119 ▲ 120 ▼

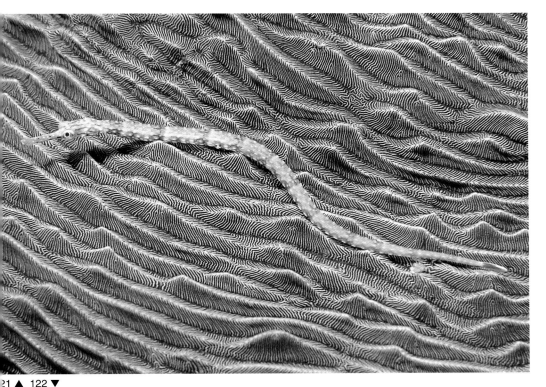

21 ▲ 122 ▼

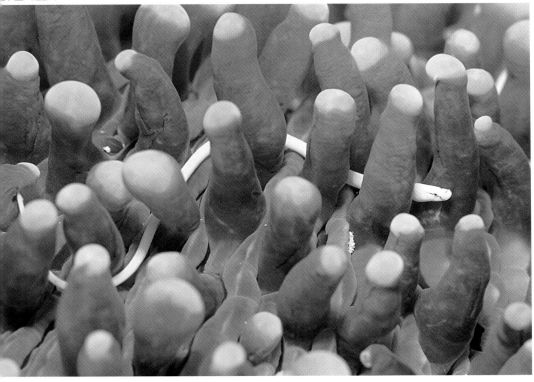

Scorpionfishes

123 ▲

124 ▲ 125 ▼

*All fishes of the scorpionfish family have venom_ dorsal fin spines. In the case of the notorious stonef_ the venom can be deadly, while in other species it is j_ agonizingly painful! Although the pain of a sting is _ tense, treatment is simple and effective: immerse the _ flicted part in hot water (at about 40-45°C, not boilin_*

123 – *The waspfishes,* **Ablabys,** *look very similar the leaffish (next page) but there are a number anatomical differences and they are sometimes plac in their own separate family.*

124 – *This diminutive shortfin scorpionfish,* **Scorpa nodes parvipinnis,** *is a nocturnal species. It can distinguished by the large pale patch behind the hea_*

125 – *The false stonefish or devil scorpionfish,* **Scorpa nopsis diabolus,** *has a characteristic humped bac_ Perhaps because it does little but sit and wait for pr_ to pass by, the false stonefish has one of the smalle_ brains of any fish.*

126 – *Scorpionfishes are usually red or brown, but oth_ colours are also seen, as in this green* **Scorpaenopsis.**

126

127 ▲

128 – *There are many species of lionfish in Indonesia, and it is not always easy to distinguish between them. This species, the spotfin lionfish,* **Pterois antennata***, has characteristic dark blotches on its pectoral fins.*

127 – *This zebra lionfish,* **Dendrochirus zebra***, is flashing its brightly patterned pectoral fins as a warning to the photographer to stay away.*

128 – *There are many species of lionfish in Indonesia, and it is not always easy to distinguish between them. This species, the spotfin lionfish,* **Pterois antennata***, has characteristic dark blotches on its pectoral fins.*

129 – *Perhaps because of its feathery appearance the lionfish,* **Pterois volitans***, is known in Indonesian as* ikan ayam, *the chicken fish.*

130 – *The leaffish,* **Taenionotus triacanthus***, comes in a variety of shades and colours, from black to white through reds, browns and yellows.*

128 ▲ 129 ▼

130 ▶

Bottom Huggers

A variety of fishes make a living right on the sea floor. As adaptations to life on the bottom, many of these have a flattened body and broad pectoral fins. Here are three of the more spectacular examples.

131 – *The seamoth,* **Eurypegasus draconis***, is a remarkably well camouflaged inhabitant of inshore sand and rubble bottoms. Its scientific name is derived in part from the name Pegasus, the winged horse of Greek mythology.*

131 ▲

132 – *The flying gurnard,* **Dactyloptena orientalis***, is a hunter of small bottom-dwelling fish and invertebrates. It is most often seen by divers on inshore reefs where it appears to be most active by night.*

133 & 134 – *Although growing to at least 50 cm in length, the giant flathead or crocodile fish,* **Cymbacephalus beauforti***, is superbly camouflaged. Unlike the previous two species, which move over the bottom in the hope of flushing out their prey, this species is content to wait until its prey passes within snapping distance. The juvenile flathead (left) is making a very good impersonation of a dead stick.*

132 ▼ 133 ▲

134 ▶

Groupers

Groupers are much sought after by fishermen through out Indonesia. The main export markets are Hong Kong and Taiwan. In some areas it is now unusual to see any groupers at all.

135 – Groupers, such as this camouflage grouper **Epinephelus polyphekadion**, are consummate predators of small fishes and other reef animals.

136 – The blacktip grouper, **Epinephelus fasciatus** is one of the most abundant groupers on Indonesian coral reefs. It is sometimes known as the redbanded grouper, but the pattern of bands on the side can be switched on and off, so this is not an ideal name.

137 – Most groupers are called kerapu or garupa in Indonesian, but the distinctive lyretail grouper **Variola albimarginata**, is known as sunu.

138 – An inhabitant of reef caves, the six-bar rockcod **Cephalopholis sexmaculata**, swims with its belly towards the nearest substrate. It is as happy swimming upside down under the cave roof as it is the right way up on the cave floor.

135 ▲   136 ▼

137 ▲ 138 ▼

Fairy Basslets

The tiny fairy basslets, sometimes known simply as anthias, are relatives of the groupers. They are plankton feeders, living in schools on reef drop-offs and other current-swept areas. They change sex as they grow, starting their adult lives as females, and later changing to become males. Many species form harems, with one male having several females under his control.

*139 – The male scalefin fairy basslet, **Pseudanthias squamipinnis**, is tinged with red and has a greatly elongated dorsal spine. The females are orange.*

*140 – The mirror fairy basslet, **Pseudanthias pleurotaenia**, lives on steep reef drop-offs, normally in depths greater than about 20 metres.*

*141 – The red-cheeked fairy basslet, **Pseudanthias hutchii**, is a common species in Indonesia. It lives on rich coral reef edges and the tops of drop-offs.*

*142 – The male purple fairy basslet, **Pseudanthias tuka**, has a distinctly pointed upper lip and a yellow chin. These features are absent in the smaller females, which boast yellow stripes along the back and tail fin.*

139 ▲ 140 ▼

141 ▲ 142 ▼

Dottybacks and Jawfishes

Dottybacks are small, usually brightly coloured reef fishes. Jawfishes live in burrows on sand and rubble. Both have a single dorsal fin, and both guard their eggs until they hatch. Dottybacks produce a ball of eggs, which the males look after in a coral crevice. Jawfishes go one better and hold their eggs in their mouths.

143 ▲

144 ▲ 145 ▼

143 – *The royal dottyback,* **Pseudochromis paccagnellae***, is one of the most strikingly coloured of all reef fishes. The white line between the yellow and magenta halves is very variable, but appears to be most developed in some Indonesian individuals.*

144 – *The purple dottyback,* **Pseudochromis porphyreus***, is a western Pacific species. In Indonesia it is only known from Maluku and Irian Jaya.*

145 – *The shy comet,* **Calloplesiops altivelis***, lives in small caves on reef slopes. It was first described from Pulau Nias off west Sumatra.*

146 – *There are many species of jawfishes in Indonesia. Several, including this species of* **Opistognathus***, have not yet been given scientific names.*

146

Snappers

Snappers are important food fishes in Indonesia a elsewhere throughout the tropics. Tens of thousand of tonnes are landed annually from Indonesian core reefs. Many snapper species are nocturnal, so they ar most easily caught using hook and line at night. B day they tend to form resting schools on the reef.

147 – Fusiliers are close relatives of the snapper They are daytime plankton feeders. This is the yellou tailed fusilier, **Caesio teres**, known in Indonesian a pisang-pisang.

148 – The two-spot red snapper, **Lutjanus bohar**, a popular edible fish, although some large specimen have caused ciguatera poisoning.

149 – The blacktail snapper, **Lutjanus fulvus**, found on coastal and lagoon reefs. It is not a larg species, rarely exceeding 30 cm in length.

150 – Adult Russell's snappers, **Lutjanus russell** are found on coral reefs. Juveniles, which are strongl striped, are found in mangrove estuaries and some times enter fresh water.

147 ▲ 148 ▼

124

Sweetlips and Goatfishes

*Sweetlips and goatfishes are all carnivorous fishe[s]
feeding on a wide variety of bottom dwelling inver[-]
tebrate animals. Goatfishes feed mainly by day, whil[e]
sweetlips feed mainly at night.*

151 – *Juvenile sweetlips have colour patterns th[at]
are very different from those of the adults. Some, suc[h]
as this juvenile spotted sweetlips,* **Plectorhinchu[s]
chaetodonoides**, *appear to mimic poisonous fla[t]
worms (page 64), swimming with vigorous hea[d]
down flapping.*

152 – *The yellow-ribbon sweetlips,* **Plectorhinchu[s]
polytaenia**. *Sweetlips are nocturnal, and this speci[es]
often rests by day under coral overhangs on the ree[f]*

153 – *This oblique-lined sweetlips,* **Plectorhinchu[s]
lineatus**, *was one of a school that regularly loiters b[y]
day near the shipwreck at Tulamben, Bali.*

154 – *Goatfishes get their name from their goat-lik[e]
'beard'. This is in fact two sensory barbels, which th[e]
goatfish uses to hunt out food in the sand. This speci[es]
is the doublebar goatfish,* **Parupeneus bifasciatus**

151 ▲ 152 ▼

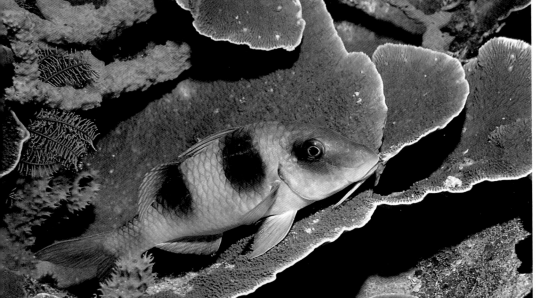

Miscellaneous Fishes

There may be over 2,000 species of fish living on cora reefs in Indonesia, and many more in the country a a whole. Nearly 100 fish families are represented i some reef areas. In a book of this size it is therefor impossible to provide more than a glimpse of th astounding diversity of Indonesian fish life.

155 – *The longnose hawkfish,* **Oxycirrhites typus** *uses a variety of perches, but is most often seen i black coral bushes.*

156 – *Like a hawk for which it was named, Forster hawkfish,* **Paracirrhites forsteri**, *sits and waits for suitable victim to pass within striking distance.*

157 – *These shrimpfish,* **Centriscus scutatus**, *liv head-down among corals or weeds. They feed on tin shrimps and other crustaceans. Shrimpfishes are als called razorfishes because of their body shape.*

158 – *Juvenile batfish,* **Platax teira**, *often form sma schools in the shelter of jetties or mooring buoys. I the background is a school of silversides (famil Atherinidae).*

155 ▲ 156 ▼

128

57 ▲ 158 ▼

159 – Sandperches, or grubfishes, are commo inhabitants of coral reefs. Most feed on crabs, shrimp and other small bottom-dwelling animals. This species the red spotted sandperch, **Parapercis schauinsland** also swims up into the water column to catch planktor

160 – Several species of flatfish live on sand and mu flats close to coral reefs. But the three-spot flounde **Samariscus triocellatus**, is a true reef dweller. It ha a peculiar habit of waving its pectoral fin in an ant clockwise direction over its head.

161 – The jacks (also known as trevallies) are maste hunters of the reef. They frequently hunt in packs patrolling the reef edge on the lookout for unwar smaller fishes. Their passage is marked by a Mexica wave of small plankton-eating fish diving for cove among the corals. Most jacks are silver, but th species, the orange spotted trevally, **Carangoide bajad**, can adopt bright yellow colouration.

162 – The mouth mackerel, **Rastrelliger kanagurta** is a filter feeder. It swims with its mouth wide oper taking in large quantities of water. Edible plankto are filtered off with special sieving apparatus on th inside of its gills.

159 ▲ 160 ▼

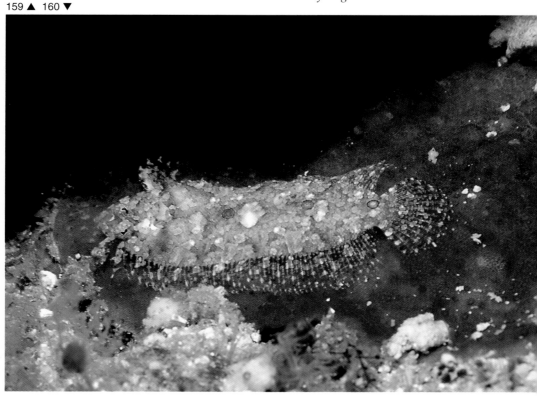

61 ▲ 162 ▼

Butterflyfishes

Butterflyfishes are emblematic fishes of the coral reef. Most species form pairs and are territorial. Their characteristic colours are yellow, white and black, usually including a dark eye band. Nearly 50 different species are known from Indonesia.

163 – The spot-tail butterflyfish, **Chaetodon ocellicaudus**, is only known from eastern Indonesia and Papua New Guinea. This individual was photographed off Flores.

164 – The threadfin butterflyfish, **Chaetodon auriga** is one of the most widely distributed of all butterflyfishes, ranging from the Red Sea and East Africa through Indonesia to Polynesia.

165 – The long-nose butterflyfish, **Forcipiger flavissimus**, uses its long tweezer-like snout to pick out tasty morsels from cracks and crevices in the reef.

166 – Bennett's butterflyfish, **Chaetodon bennetti** might be confused with other yellow butterflyfishes with black spots. It can be distinguished by the pair of blue-white lines crossing its lower abdomen.

163 ▲

164 ▲ 165 ▼

166 ▶

167 ▲

167 – *The saddled butterflyfish,* **Chaetodon ephip pium***, is one of the most striking of all coral ree fishes. Butterflyfishes are called 'kepe-kepe' or 'ikar kupu-kupu' in Indonesian.*

168 – *This is the Pacific form of the red-fin butterfly fish,* **Chaetodon trifasciatus***, photographed ir Maumere Bay, Flores. It has different colouratior from the Indian Ocean form and is sometimes classifiec as a separate subspecies, or even as a separate species* **Chaetodon lunulatus***.*

169 – *Pairs or small groups of eye-patch butterflyfishes* **Chaetodon adiergastos***, are usually seen resting near a coral outcrop.*

170 – *The spot-banded butterflyfish,* **Chaetodor punctatofasciatus***, is found on steep, coral-rich outer reefs.*

168 ▲ 169 ▼

170 ▶

171 ▲

Angelfishes

Angelfishes are very closely related to butterflyfishes. In the field they can be distinguished by the presence of a strong cheek spine, which is absent in butterflyfishes. Large angelfishes can produce clearly audible grunting or thumping noises when disturbed, which can sometimes startle unsuspecting divers.

171, 172 & 173 – Among large angelfishes of the genus **Pomacanthus** the juveniles have a very different colour pattern from the adults. This is believed to reduce aggression by adults against youngsters. As an example, juveniles of the imperial angelfish, **Pomacanthus imperator**, have white lines on a blue background, while adults are much more colourful (small juvenile above, juvenile centre, adult below).

174 – Because of its exquisite colouration the regal angelfish, **Pygoplites diacanthus**, is a popular aquarium fish. It is, however, a fussy feeder and very difficult to maintain in captivity.

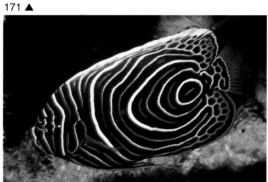

172 ▲ 173 ▼

174 ▶

175 ▲

176 ▲ 177 ▼

175 – *Angelfishes of the genus* **Centropyge**, *like this keyhole angelfish,* **Centropyge tibicen**, *are algal grazers. They are very timid, and dive into reef crevices at the first sign of danger.*

176 – *Eibl's angelfish,* **Centropyge eibli**, *is named in honour of biologist Irenäus Eibl-Eibelsfeldt, who accompanied diving pioneer Hans Hass on several of his early diving expeditions.*

177 – *The blue-ringed angelfish,* **Pomacanthus annularis**, *like other* **Pomacanthus** *angelfishes, feeds heavily on sponges. Very few other animals are able to do this, because the sponges are heavily defended with both spiky skeletal rods and toxic chemicals.*

178 – *This three-spot angelfish,* **Apolemichthys trimaculatus**, *clearly shows the strong cheek spine that is a characteristic of angelfishes.*

178 ▶

Damselfishes and Anemonefishes

There are more species of damselfish (about 140 known so far) found in Indonesia than in any other country. Damselfishes are also the most abundant group of fishes on Indonesian coral reefs, in terms of numbers of individual fish. Their importance to reef ecology is immense.

179 – The anemonefishes are a specialized damselfish subfamily. Thirteen of the 28 species of anemonefish and all 10 species of host anemone are found in Indonesia. This is a pair of orange anemonefishes, **Amphiprion sandracinos**, with their host sea anemone **Stichodactyla mertensii**.

180 – The clownfish, **Amphiprion ocellaris**, has a distinctive middle bar, shaped like a pair of underpants.

181 – The sergeant major, **Abudefduf vaigiensis**, is widely distributed throughout the Indo-Pacific. However, it is named after the locality where it was first collected, Waigeo Island off Irian Jaya.

182 – Female blue devils, **Chrysiptera cyanea**, as shown here, have a black spot near the back of the dorsal fin. Males lack the spot, and in some areas have an orange tail.

179 ▲   180 ▼

Wrasses

The wrasses are a particularly diverse family of fishes which range in size from several tiny species, just a few centimetres long when fully grown, to the massive napoleon (or humphead maori) wrasse, which can grow to well over two metres long. Many species change sex as they grow, starting their adult lives as females and becoming males at a larger size.

183 – Rockmover wrasses, **Novaculichthys taeniourus**, are named for their habit of turning over rocks and pebbles in search of invertebrate prey.

184 – This moon wrasse, **Thalassoma lunare**, is a mature male, resplendent in nuptial blue.

185 – This black belt hogfish, **Bodianus mesothorax**, was photographed on the rich coral reefs of the Marine Park (Taman Nasional Laut Bunaken-Manado Tua) off Manado, Sulawesi.

186 – The checkerboard wrasse, **Halichoeres hortulanus**, is a familiar sight on reefs throughout much of the Indo-Pacific.

183 ▲ 184 ▼

185 ▲ 186 ▼

Parrotfishes

Parrotfishes are so-called because of their brigh colouration and their powerful parrot-like beaks. The use their beaks to graze on seaweeds. These of cours require sunlight to grow, so parrotfishes are see much more commonly in shallow than in deep wate

187 – *The juvenile bicolour parrotfish,* **Cetoscaru bicolor**, *has a unique colour pattern which makes look more like an anemonefish than a parrotfish.*

188 – *The spectacular humphead parrotfish* **Bolbometopon muricatum**, *can grow to well over one metre in length. Schools of humpheads graze lik herds of buffalo on the reef, consuming prodigiou quantities of coral and algae.*

189 – *The yellow-head or pygmy parrotfish,* **Scaru spinus**, *is a small species, reaching little more tha 30 cm in length.*

190 – *Parrotfishes are active by day, and sleep i hollows in the reef at night. This sleeping Singapor parrotfish,* **Scarus prasiognathus**, *was photographe at night off Ambon.*

187 ▲

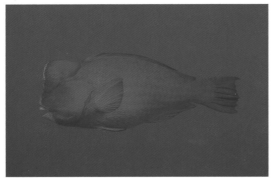
188 ▲ 189 ▼

190

Blennies and Triplefins

Blennies are often mistaken for gobies (next page) Blennies can be distinguished by their blunt face adorned with cirri (sensory tufts), and their lon single dorsal fin (usually separated into two in gobies

191 ▲

191 – *The startling light blue markings on the face this black comb-toothed blenny,* **Ecsenius namiye** *may be a fright pattern.*

192 – *The pinstripe blenny,* **Ecsenius pictus***, no mally occurs on coral reefs in depths of 10-40 metre.*

193 – *This beautiful blenny,* **Ecsenius melarchus***, only known from central Indonesia, Sabah and souther Philippines. This individual was photographed o East Kalimantan. In some other areas the iris is brig yellow, giving it the name yellow-eyed blenny.*

194 – *Triplefins are small fishes closely related blennies. As the name suggests, they have thr dorsal fins, instead of the one or two found in mo other fishes. This striped triplefin,* **Helcogramm striata***, is probably the commonest member of i family on Indonesian reefs.*

192 ▲ 193 ▼

194

Gobies

195 ▲

196 ▲ 197 ▼

There are hundreds of species of goby in Indonesia, far more than of any other fish family. Most are small, and many are rather dull in colour or retiring in habit, so they are easily overlooked. But look closely: there are many fascinating and beautiful species.

195 – *While sandy bottoms may appear lifeless to the inexperienced diver, they are almost always home to a variety of delightful animals, such as this black-rayed shrimp-goby,* **Stonogobiops nematodes.**

196 – *The crab-eyed goby,* **Signigobius biocellatus** *raises its fins when alarmed in apparent mimicry of a belligerent crab.*

197 – *Several species of goby are only found in association with particular reef invertebrate hosts. For example, this sea whip goby,* **Bryaninops yongei***, is invariably associated with black coral whips,* **Cirripathes***.*

198 – *Several gobies, such as this pink coral goby,* **Pleurosicya mossambica***, live in symbiotic association with soft corals.*

198 ▶

Surgeonfishes and Rabbitfishes

Surgeonfishes and rabbitfishes are both herbivorous algal grazers, and both often form feeding schools. In addition, both are armed with potentially dangerous spines. If stung, pain can be severe but will normally fade without treatment within a couple of hours.

199 – *The black mark on the side of this blue surgeonfish,* **Paracanthurus hepatus**, *gives it its alternative English name (palette surgeonfish) and its Indonesian name ('ikan angka enam', number six fish).*

200 – *This sailfin tang,* **Zebrasoma veliferum** *was photographed on the famous wreck at Tulamben in Bali. This is a Pacific species, found in central and eastern Indonesia.*

201 – *The yellow foxfaced rabbitfish,* **Siganus vulpinus***, lives in pairs. It is found in rich coral areas throughout Indonesia.*

202 – *Unlike most rabbitfishes, which are active by day, the gold-saddled rabbitfish,* **Siganus guttatus***, appears to be active at night, spending much of the daytime loitering near coral outcrops or on sandy patches.*

199 ▲ 200 ▼

201 ▲ 202 ▼

151

Triggerfishes and Filefishes

Triggerfishes are named for the structure of their dorsal fins. The long first dorsal spine can be locked into an upright position. This enables the triggerfish to wedge itself into crevices, to sleep or when threatened. Pressure on the small second dorsal spine (the trigger) unlocks the first spine, allowing the fish to be removed. Most of the time the spiny dorsal fin is folded down on the back.

203 – Clown triggerfishes, **Balistoides conspicillum,** *are highly prized in the aquarium trade.*

204 – As the name suggests, lagoon triggerfishes, **Pseudobalistes flavimarginatus,** *are found mainly in sandy areas.*

205 – Filefishes are related to triggerfishes, differing in fin structure and dentition, and in having a prickly or furry skin. This unfortunate filefish, **Acreichthys tomentosum,** *has two parasitic sea lice on its back.*

206 – One of the most respected of all reef fishes is the titan triggerfish, **Balistoides viridescens.** *It lays its eggs in a nest, which it will defend vigorously against all comers. If you are attacked, retreat is the sensible option!*

203 ▲

204 ▲ 205 ▼

206 ▶

Pufferfishes and Relatives

As the name suggests, pufferfishes are able to puff themselves up, a trick they perform by taking in water through the mouth. This makes them much bigger than normal, and so almost impossible for would-be predators to swallow. They also pack a powerful poison in their skin and internal organs.

207 – Boxfishes are related to pufferfishes, but have a tough, rigid body rather than an inflatable one. This is the yellow boxfish, **Ostracion cubicus**.

208 – The black-spotted puffer, **Arothron nigropunctatus**, is perhaps the commonest pufferfish on Indonesian reefs. Its colour can vary from brown to grey.

209 – Large pufferfishes have very strong jaws. This adult scribbled pufferfish, **Arothron mappa**, is being attended by a cleaner wrasse, **Labroides dimidiatus**.

210 – Porcupinefishes, like this black-blotched **Diodon liturosus** are able to inflate themselves just like pufferfishes. However, this defence mechanism is reinforced by numerous stout spines, which become erect as the fish puffs up.

207 ▲ 208 ▼

154

Marine Reptiles

Turtles, sea snakes and the saltwater crocodile are *the* marine reptiles. They all have scaly skins, breathe air and in most cases return to the land to lay their eggs. Exceptions to the latter are those sea snakes that produce live young. Sea snakes are highly venomous and often inquisitive, but they are not dangerous to divers or snorkellers. Salt water crocodiles on the other hand have a bad reputation that is wholly deserved. Fortunately they are rarely seen on coral reefs, preferring estuaries and remote mangrove areas.

211 – *The banded sea snake,* **Laticauda colubrina**, *can be distinguished from similar banded species by its yellow upper lip. Pulau Manuk and Gunung Api in the Laut Banda are two volcanic islands with especially large sea snake populations.*

212 – *Green turtles,* **Chelonia mydas**, *are much hunted. Many end their days in Bali, where they are used in religious ceremonies, as well as for food.*

213 – *Turtle tracks on the beach at Sangalaki, East Kalimantan. Without nesting sanctuaries such as this, the chances for turtles to survive in Indonesia would be greatly diminished.*

211 ▲ 212 ▼

213

INDEX

158

Selected Reading

Allen G.R. and R. Steene. 1994. *Indo-Pacific Coral Reef Field Guide*. Tropical Reef Research, Singapore. 378 pp.

Bleeker P. 1877. *Atlas Ichthyologique des Indes Orientales Nederlandaises*. 9 volumes. Amsterdam. (US reprint available).

Colin P.L. and C. Arneson. 1995. *Tropical Pacific Invertebrates*. Coral Reef Press, Beverly Hills. 296 pp.

Dharma B. 1988 & 1992. *Siput dan Kerang Indonesia* (Indonesian Shells). Volumes 1 & 2. Christa Hemmen Verlag, Germany. 111 pp & 135 pp.

Fautin D.G. and G.R. Allen. 1997. *Anemonefishes and Their Host Sea Anemones*. Second, revised edition. Western Australian Museum, Perth. 160 pp.

Gosliner T.M., D.W. Behrens and G.C. Williams. 1996. *Coral Reef Animals of the Indo-Pacific*. Sea Challengers, Monterey. 314 pp.

Kuiter R.H. 1992. *Tropical Reef-fishes of the Western Pacific: Indonesia and Adjacent Waters*. PT Gramedia Pustaka Utama, Jakarta. 314 pp.

Kuiter R.H. and H. Debelius. 1994. *Southeast Asia Tropical Fish Guide*. IKAN, Frankfurt. 321 pp.

Ming C.L and P.M. Alino. 1992. *An Underwater Guide to the South China Sea*. Times Editions, Singapore. 144 pp.

Muller K. 1997. *Diving Indonesia: a Guide to the World's Greatest Diving*. Revised, third edition. Periplus Editions, Singapore. 332 pp.

Severns M. 1996. *Sulawesi Seas: Indonesia's Magnificent Underwater Realm*. Periplus Editions, Singapore. 160 pp.

Stafford-Dietsch J. 1996. *Mangrove: the Forgotten Habitat*. Immel Publishing, London. 277 pp.

Tomascik T., A.J. Mah, A. Nontji and M.K. Moosa. 1997. *The Ecology of the Indonesian Seas, Parts One & Two*. (The Ecology of Indonesia Series, Volumes VII and VIII). Periplus Editions, Singapore. Pp. 1-642 & 643-1388.

Veron J.E.N. 1986. *Corals of Australia and the Indo-Pacific*. Angus and Robertson, Sydney. 644 pp.

Wallace A.R. 1867. *The Malay Archipelago: the Land of the Orang-Utan and the Bird of Paradise: a Narrative of Travel, with Studies of Man and Nature*. London. 653 pp.

Wilkinson C.R., S. Sudara and C.L. Ming (editors). 1994. *Proceedings of the Third ASEAN-Australia Symposium on Living Coastal Resources. Volume 1: Status Reviews*. AIMS, Townsville. 454 pp.